I0048485

THE HOP

ITS CULTURE AND CURE, MARKETING AND MANUFACTURE

A practical handbook on the most approved methods in growing, harvesting, curing and selling hops, and on the use and manufacture of hops

BY

HERBERT MYRICK

Published by Left of Brain Books

Copyright © 2021 Left of Brain Books

ISBN 978-1-396-32219-8

First Edition

All rights reserved. No part of this publication may be reproduced, distributed, or transmitted in any form or by any means, including photocopying, recording, or other electronic or mechanical methods, without the prior written permission of the publisher, except in the case of brief quotations embodied in critical reviews and certain other noncommercial uses permitted by copyright law. Left of Brain Books is a division of Left of Brain Onboarding Pty Ltd.

Table of Contents

ACKNOWLEDGMENTS   1

CHAPTER I. ORIGIN AND SPREAD OF HOP CULTURE   3

   Foreign Hop Plantations   4

   Where Hops are Grown in America   8

   Future of the Hop Industry   12

CHAPTER II. PECULIARITIES OF THE INDUSTRY   14

   Uses of the Hop   19

CHAPTER III. CHARACTERISTICS OF THE PLANT   22

   Seedless Versus Seed Hops   29

   Varieties of Hops   33

   What Constitutes Quality in Hops?   41

   Brewers' Views in Buying Hops.   44

CHAPTER IV. COMPOSITION OF THE HOP PLANT AND ITS FRUIT   47

   The Chemistry of Hops   49

CHAPTER V. THE CLIMATE AND SOIL FOR HOPS   54

   The Best Soil for Hops   55

   Location of a Hop Yard   57

   Preparation of the Soil   59

CHAPTER VI. FEEDING THE HOP PLANT   61

CHAPTER VII. LAYING OUT A YARD—TRAINING THE VINES   71

CHAPTER VIII. PLANTING AND CULTURE   79

   Cultivation During the Second Year   84

   Tying Up the Vines   90

CHAPTER IX. METHODS OF THE PLEASANTON HOP COMPANY   93

CHAPTER X. PESTS OF THE HOP CROP   100

   The Hop Plant Louse (*Phordon humuli*, Schrank)   102

The Hop Grub or Hop-Plant Borer (*Gortyna immanis*, Grt.)   112

Caterpillars Feeding upon Hop Leaves   114

The Hop Vine Snout-Moth (*Hypena humuli*, Harr.)   114

Hop Merchants (*Polygonia interrogationis*, Godart, and *Polygonia comma*, Harr.)   115

The Zebra Caterpillar (*Mamestra picta*, Harr.)   118

The Common Woolly Bear Caterpillar (*Spilosoma virginica*, Fab.)   119

The Saddle-Back Caterpillar (*Empretia stimulea*, Clem.)   120

Other Caterpillars   121

Leaf Hoppers which affect the hop   121

Beetles Feeding on Hop Leaves   123

The So-Called "Red Spider," or Spinning Mite   124

Practical Directions for Spraying   125

Fungus Pests—Blight, Molds, etc.   131

Other Pests   139

CHAPTER XI. HARVESTING THE CROP   141

Pickers and prices   145

Rules for picking and pickers   146

CHAPTER XII. KILNS FOR CURING HOPS   156

English Oast Houses   157

Hop Kilns in New York State   159

Hop Kilns on the Pacific Coast   166

CHAPTER XIII. CURING, COOLING AND BALING HOPS   170

Curing the Hops   170

Cooling and Baling   176

Additional Notes on Curing   185

The Sulphuring of Hops   191

CHAPTER XIV. GRADING AND MARKETING HOPS   196

Sampling Hops Preparatory to Selling   201

Marketing the Crop   206

Storing Hops for Long Keeping   210

Extracting the Lupulin   212

CHAPTER XV. CONCENTRATION IN HOP GROWING   216

CHAPTER XVI. EXPENSES AND PROFITS OF HOP CULTURE   222

Cost of Hops in New York State, U.S.A.   224

New York State—Cost of Growing Hops   225

From Northern Ohio   229

Cost of Hops on the Pacific Coast   231

Washington, King County, 1897 Crop (By Alexander Adair).   234

Raising the Crop in the Northwest   234

APPENDIX   236

Statistics of the Hop Trade   236

THE HOP DICTIONARY   250

The Hop Glossary   252

RULES REGULATING THE HOP TRADE   271

BIBLIOGRAPHY   274

ACKNOWLEDGMENTS

In the preparation of this work the author has had the co-operation of many of the leading hop growers and dealers in two continents, to whom his debt is most gratefully acknowledged for facts, experiences and photographs. The United States department of agriculture has furnished certain illustrations, while official statistics and returns have been supplied by the United States treasury department, the English board of agriculture and the German foreign office. The somewhat scanty literature on the subject has been freely drawn upon, including nearly all the works listed in our bibliography. During the past fifteen years that the author has been collecting data on this subject, in connection with "Our Hop Growers' Exchange" department in *American Agriculturist,* many invaluable statements have been received and these have also been fully employed, particularly numerous essays of practical men on the cost of growing hops. Special services have been rendered that should have special recognition.

From the state of Washington came important helps by E. Meeker and James Hart, also Major R. M. Hornsby of British Columbia. Oregon: A. J. Wolcott, H. J. Ottenheimer. California: Lilienthal & Co. of San Francisco, and the Pleasanton Hop Company afforded every possible assistance in the way of photographs, statistics of coast crops, etc.; Daniel Flint, the hop pioneer, was a liberal contributor from his experience; also L. F. Long and others, while Horst Brothers placed at our disposal all the experience and resources of their various hop plantations. New York: James F. Clark, the largest hop grower in the state; W. A. Lawrence, Editor W. S. Hawkins of the Waterville Times, Secretary Fox of the New York City hop trade.

In England, the London hop dealers extended every assistance, also numerous growers. All the results of the scientific experiments conducted at the Southeastern agricultural college at Wye, in Kent, were generously made available for this work by President Hall. Editor E. H. Elvy of the *Kentish Observer* aided with valuable data and pictures. Editor Ironmonger's work in the English *Hop Grower* (a useful journal, now defunct), has also been an important aid, and he has contributed otherwise to this book. As secretary of the National Association of English Hop Growers, Mr. Thomas Ironmonger has also rendered much valuable assistance. Special credit should be given to Charles Whitehead's works.

In Europe, we are under special obligations to C. Beckenhaupt of Alsace, Von Barth & Co., the Nuremberg merchants, Editor Fairt of the *Deutschen Hopfenbau Verein*, and many others.

Dr. L. O. Howard, chief of the division of entomology of the United States department of agriculture, prepared the most of the admirable chapter on hop insects. Dr. H. W. Wiley, chief of the division of chemistry, aided in preparing the chemistry of the hop plant, as presented by E. E. Ewell, assistant in that division. N. F. Walter's glossary of hop terms is a distinct contribution to technical literature. C. F. Dalton deserves much credit for assistance in putting the book to press.

In all modesty, therefore, this book may be accepted as a comprehensive treatise on its special topic. Particular pains have been taken to make it strictly accurate, so that it may be the authority upon all points pertaining to hops of which it treats.

CHAPTER I.
ORIGIN AND SPREAD OF HOP CULTURE

FOR more than a thousand years the virtues of hops have been recorded, and this remarkable plant has doubtless been cultivated since almost prehistoric times. Certainly more than 500 years have gone since domesticated hops were brewed in middle Europe, but the wild hops were used much earlier, and are brewed in Styria to the present day. Long before Columbus sailed the pathless sea, the wild hop was well established in England, but came into prominence only after its culture was introduced into Kent by Flemish immigrants about 1524.

Though this plant grows luxuriantly throughout the temperate regions, such are its peculiarities that the commercial crop has been confined to nearly the same localities in England and Europe from earliest times. Kentish growers have held first place ever since Parliament legalized this industry in 1554, and while the area under hops in England has fluctuated materially during the past century, the crop has been confined to essentially the same districts. America has witnessed the same tendency of commercial hop growing to concentrate. Though introduced into the New Netherlands in 1629, and into Virginia about 1648, and encouraged by special legislation until 1657, hop culture has assumed importance in the United States only since 1800. As the industry developed, it centered in New York state, though many hops were grown in Wisconsin after the Civil war, but of late years certain districts on the Pacific coast have proven to be so adapted to this crop as to seriously threaten the older established hop yards on both sides of the Atlantic.

Hops are raised for family and medicinal purposes in other states of America and other countries, but the commercial crop is now nearly confined to certain limited sections, as it has been for many years, the modern development on the Pacific coast excepted. The statistical tables in the appendix show that the average area devoted to this crop during the closing decade of the century may be thus roughly stated:

Foreign.	Acres.	American.	Acres.
Germany	100,000	New York	25,000
England	55,000	California	7,000
Austria	37,000	Oregon	15,000
France	7,000	Washington	6,500
Other	11,000	Other	1,000
Total	**210,000**		**54,500**

World's average aggregate 265,000 acres.

FOREIGN HOP PLANTATIONS

Germany—Although this country produces the bulk of the hops grown on the continent, the number of large plantations is limited. The hops are grown usually in comparatively small fields, and in many cases in small garden patches. Hops are raised by the German family as a side issue, much as the American farmer's family raises poultry. The culture is largely by hand, and its special features are embodied in subsequent chapters on methods of culture, along with methods used by English and American growers. Even in Bavaria, the principal hop-producing section of Germany, the hop yards will not average much over one or two acres in extent. The picking is done by the family; in bad weather, the vines are cut and taken indoors to be picked. The growers do not have curing houses, but sell the sun-dried hops to the dealer, who attends to the proper curing and sulphuring. The bulk of Germany's crop is produced in Bavaria, which furnishes one-half or more of the empire's product. Then follow in order of importance—Wurtemberg, Baden, Posen, Altmark and scattering districts. Nuremberg in Bavaria is the controlling market for German hops, although hops are bought by dealers direct from growers at many other points.

The *French* district is largely confined to Alsace and Lorraine, now German provinces, but hops are grown to a considerable extent in Burgundy and Northern France. The industry is decreasing in this section.

Training the Hoppe.

"It shalle not be amisse nowe and then to passe through your Garden, having in eche hande a fothed wande, directyng aright such hopes as declyne from the poales."

Gathering the Hoppe.

"Cutte them" (the hop stalkes) "a sunder wyth a sharpe hooke, and wyth a fothed staffe take them from the poales."

5

Of Ramming of Poales.

"Then with a peece of woode as bigge belowe as the great ende of one of youre poales, ramme the earth that lieth at the outsyde of the poale."

Cutting Hoppe Rootes.

"When you pull downe your hylles...you should undermine them round about."

Of Tying of Hoppes to the Poales.

"When your hopes are growne about one or two foote high, bende up (with a ruthe or a grasse) such as decline from the poales, wynding them as often about the same poales as you can, and directing them always according to the course of the Sunne."

In *Austria-Hungary,* attention to hop culture is increasing slightly. Special grades of hops, with peculiar characteristics that give them a fancy value, are grown in upper Austria, including Galicia, Styria, Silesia and Moravia, also further south in Hungary. But the center of the Austro-Hungarian industry is in Bohemia, where between 25,000 and 30,000 acres are usually devoted to hops. One-half of the acreage is located in the Saatz district, the hops from which command the highest prices in the world's market—two and three times as much as American or English hops. The quality of Bohemian hops is carefully safeguarded by government, which has established technical schools in hop culture at Rakonitz and Laun. Besides hop gardens and laboratories for scientific work, these schools are provided with an elaborate course of instruction and experimentation. So interesting and useful is the study that students go to these schools from other countries. Every bale of hops produced in Bohemia must be officially sealed by a government inspector, which insures hops of true grade in that country, but does not prevent the adulteration of Bohemian hops when exported.

Elsewhere.—In Holland and Belgium, the acreage devoted to hops has been reduced until these crops no longer have much influence on market values, although several thousand acres are still devoted to the crop.

In Russia, about 8000 acres of hops were formerly grown in Kieff and Volhynia, but owing to a heavy reduction in duty and other causes, the commercial area has been reduced, but efforts are again being made to widely increase the industry in Russia. The scattered hop fields in other parts of Europe are too insignificant to be mentioned.

Australasia has for years had less than 2500 acres devoted to this crop, but it is believed that the industry is capable of large development in that country.

In *England*, about two-thirds of the usual hop area is confined to Kent, the other counties being in order of importance—Hereford, Sussex, Worcester, Hants and Surrey. Following the period of high prices, the English crop reached a total extent of 70,000 acres in 1886, but has steadily declined to around 50,000 acres during the closing years of the century.

WHERE HOPS ARE GROWN IN AMERICA

New York State.—In 1808, the first yard was set out in the state of New York by James D. Coolidge at Madison. The demand was gradually increasing, and the area planted to hop yards or small plantations was slowly extended where the conditions of climate and soil seemed favorable. The quality of the American hop was considered by the brewers in those days as very inferior, and the prices paid for them were much below those of English hops that were imported. There were also difficulties in delivering the crops to market, as they had to be hauled long distances by teams of oxen. The heavy crops grown, 2000 lbs to the acre, proved profitable even at the low prices then obtained, about 10 or 12 cents per pound.

A succession of bad crops in England, however, stimulated the industry, especially in New York state, where the soil in some sections had been particularly adapted to hops. The first actual statistics of the hop crop of the United States were for the year 1840, when the total crop was estimated at 6200 bales of 200 lbs., or a total harvest of 1,240,000 lbs., of which two-thirds were grown in the New England states and one-third in New York. During the next ten years the hop industry nearly trebled in extent, the entire crop of the country being 3,497,000 lbs. in 1849, or 17,500 bales, of which New York state raised five-sevenths, New England producing only a little more than 700,000 lbs. that year, with scattering lots in Illinois, Indiana, Michigan, Missouri, Pennsylvania and Wisconsin.

During the ensuing decade the hop crop of the United States again trebled in quantity, the total yield in 1859 being about 55,000 bales, of which New York state grew seven-eighths, Vermont being the only New England state to "stay in the ring." Between 1860 and 1870, the increase was smaller, 150 per cent., the crop of 1869 amounting to 127,000 bales, of which New York state produced 97,500 bales. Western competition, low prices and poor crops now conspired to reduce New York's hop area until the federal census of 1890 showed that this state produced only about half the nation's hops. The proportion of America's crop now grown in the Empire state is still less, and the future will show whether this crop, like so many others, is to go entirely west. The principal hop counties of New York state have stood for years in this order of importance: Otsego, Madison, Oneida, Schoharie, Franklin, Montgomery, Ontario.

In New York state, dairying and hop growing are generally combined, the manure from the cattle being needed to fertilize the hop roots. Hop growing often proves a failure with small growers, owing partly to disease and parasites and partly to low prices. The small grower also is occupied with other crops and has not time to give as much care and attention to the hop yards as they deserve, the plant being prompt to resent any neglect. It is in the small yards that lack of cultivation is so common, together with carelessness in tending the crop, looking after the poles, or tying the vines. The largest yard in New York state is that of James F. Clark, whose yard, near Cooperstown, covers 150 acres, which are always brought to a high state of cultivation. Waterville, Cooperstown and Schoharie are the market centers for New York state hops.

Wisconsin embarked in hop culture in the early sixties, and by 1869 the federal census showed a crop of 25,000 bales. This has never since been equaled or exceeded. Ten years later, Wisconsin produced less than half that quantity of hops, and since then its product has steadily diminished, never exceeding half a million pounds. The crop has been reduced by lice, and comparatively few growers gave it the attention bestowed upon hops in New York. Wisconsin plantations are now confined to a few large yards of from 10 to 100 acres, less than 1,000 acres being devoted to the crop and often but a fraction of the area is worth picking.

California led off in the introduction of hop culture to the Pacific coast. Daniel Flint brought the first hops into the Golden state, in 1857, from Vermont. He persisted in their culture almost alone until the legislature of 1863 voted him $1000 as a reward for demonstrating the possibilities of this new crop in the Sacramento valley. From 8000 bales in 1869, the California crop has jumped to some 50,000 bales, grown on some 7500 acres, compared to 1100 acres in 1879. The largest hop plantations in the world are along the rich alluvial bottom lands of the Sacramento, Russian and Feather rivers in California. The size of a hop farm in that state ranges all the way from 10 to 300 acres, the latter being the size of the Pleasanton plantation, Alameda county, where at harvest time as many as 1500 to 2000 pickers are employed. The principal hop growing counties are Sonoma, Sacramento, Mendocino, Alameda, Yolo, Yuba, San Joaquin.

Oregon's commercial hop industry dates from about 1880, and has been characterized by wide fluctuations in area devoted to the crop, likewise in yield and quality. These violent changes are due partly to the fact that on these rich soils hop cuttings planted in spring will yield 800 to 1200 lbs. of cured hops in the fall, while in New York state no crop is expected until the second year, and not much until the third season from planting, while in England and on the continent, four years from planting are required for a full crop. This apparent advantage has operated to the detriment rather than to the benefit of the Pacific coast, especially in Oregon and Washington, because it has led to hop planting by inexperienced persons, or to the setting out of larger plantations than the owners could properly operate except by incurring heavy mortgages. Low prices following overproduction have therefore ruined a larger proportion of those who went into hops on the Pacific coast than in any other part of the world. The industry in Oregon is now confined to the counties west of the Cascade mountains, centering mainly in Marion, Polk, Clackamas, Yamhill and Washington counties.

In *Washington*, conditions are much similar to those in the neighboring state of Oregon. Although hops are being increasingly grown in the Yakima valley east of the Cascades, and to a very limited extent in the valley of the Columbia, Spokane and Snake rivers, the industry has long centered in King and Pierce counties, in the rich plains and valleys running down to the inland sea. Lewis county, Southern Washington, is also becoming quite a hop center. The statistics in the appendix show the marked variations that have characterized the areas and yield.

FIG. 1. A THREE HUNDRED ACRE HOP FIELD
NEARLY READY TO PICK.
This is one of the largest blocks of hops grown in one field
anywhere in the world. It is at Pleasanton, Cal.

FUTURE OF THE HOP INDUSTRY

The *World's Supply* of hops, it will be seen, comes mainly from the United States, England and Germany. Great Britain imports an average of 125,000 bales of hops yearly, of which 65,000 come from the United States and the balance from Europe. Germany exports about 130,000 bales per year, and imports some 20,000 bales; about 50,000 bales of her exports go to Great Britain, the balance to other European countries and to the United States. The limitation of the world's market for hops is therefore clearly defined.

The appendix tables show how both area and yield are fluctuating, and throw a flood of light on the possible monopoly of the world's hop market by the United States, and especially by our Pacific coast states. The author believes such monopoly to be possible, at least to the extent of the United States producing the largest share of the world's consumption. To that end, this book is written. But if the United States is to achieve that distinction, it will be by improving the quality of American hops until they are the best in the world and by producing them at less cost than they can be grown elsewhere.

The steady increase in the consumption of hops is also apparent from the statistical appendix. While the figures are not as perfect as desirable, because of the obvious difficulty of collecting full returns, they demonstrate a constant growth in the demand for hops. Substitutes and adulterants check the use of hops to a considerable extent, especially in seasons of scarcity, and constitute an evil that must be suppressed. The main reliance of the hop grower is the brewers' demand. The consumption of beer, already enormous, has increased astonishingly of late years, and bids fair to continue to do so. Throughout the world the tendency seems to be to replace the heavy beverages and injurious liquors with the lighter wines and beer. Brewing makes a product so much cheaper than wine that beer is destined to hold first place until humanity reaches the stage in its evolution that is characterized by total abstinence.

An increasing demand being thus assured, another favorable influence is the fact that the value of this crop is of late years being more governed by the law of supply and demand than formerly. The increasing efficiency of the crop reporting service, especially that conducted by *American Agriculturist* in co-operation with hop growers, has done something to mitigate the gambling that has characterized the selling of hops. Much more could be done to place the industry on a safer commercial basis, as suggested in the chapter on marketing, but it will require years of effort to educate growers up to the co-operation needed to accomplish this purpose.

In spite of the peculiarities of the plant and of the hop industry, as set forth in Chapters II and III, the hop for many years will continue to be an agricultural specialty that will yield profits according to the judgment employed in its culture and sale.

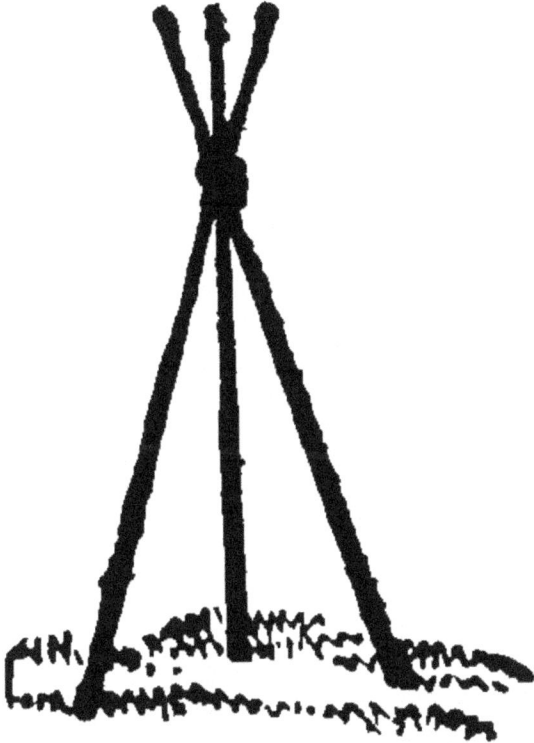

FIG. 2. COMMENCEMENT OF POLE STACK.

CHAPTER II.
PECULIARITIES OF THE INDUSTRY

THE hop industry may be regarded as a very peculiar one in many respects. The area upon which hops can be grown is limited, owing to peculiarities and necessary conditions of soil and climate, not only in this country, but throughout the world. Unfavorable weather at the critical period of hop development may almost ruin in a few days what had promised to be a crop large in quantity and fine in quality. Earlier in the season, lice and other pests may cause such injury that, even with ordinarily favorable weather, the plant may not fully recuperate and the yield will be poor.

These risks are more serious with hops than with almost any other plant. Add the dangers usual to all husbandry from drouths, wind, flood, frost, etc., and it will be seen that on a given area the product and quality of hops may vary more widely from year to year than is the case with almost any other crop. If, in addition to these conditions, the area devoted to hops should he suddenly enlarged; or, on the other hand, if considerable areas should not be harvested, owing to poor crops or low prices, wide fluctuations may occur in the supply and quality of hops. These factors make it difficult to collect exact data about the production of hops, even with the co-operation of growers. The absence of organization among growers in America, and a still worse condition in this respect in England and on the continent, adds to the uncertainty of even the best efforts to ascertain the extent of the new crop.

The effect of these natural influences that favor fluctuating supplies and prices, is heightened by artificial conditions. The movement of hops, as shown by actual shipments, and by imports and exports, fails to reveal the extent of old stocks in hands of dealers or brewers. Under ideal conditions, hops can be kept in cold storage for months without losing their virtue to any great extent. Breweries are now equipped with cold storage for this purpose, and brewers usually make it a practice to stock up liberally when prices are low, but as practiced cold storage is not proving a success. The quantity of hops used per barrel of beer varies in different breweries to such an extent as to afford but a shaky basis for computing consumption on output of beer. It is asserted by some that fewer hops per barrel of beer are now used than formerly, while others claim to the contrary. It is now customary in the American trade to estimate one pound of hops to be used for each barrel of beer, against one and one-half pounds twenty years ago, but this is somewhat arbitrary. When hops are dear, less is used per barrel than when cheap, the deficiency being made good by hop extract or substitutes.

FIG. 3. A NEW YORK HOP YARD.
Trained on long poles and cross strings.

While the supply and the uncertainty about it thus fluctuates, the demand is fairly constant in comparison. Except in so far as substitutes are used, the demand for hops is regulated by the consumption of beer, the quantity used for medicinal or household purposes really playing no part in the commercial question of supply and demand. As the consumption of beer is largest in Europe, where the population is most settled, it is not liable to sudden increases or decreases, and may be estimated with a very fair degree of accuracy year in and year out. In America, however, owing no doubt to the more sudden growth of our foreign population, the consumption of beer has increased more than in Europe. The world's steady increase in beer consumption indicates that the demand for hops is not likely to vary to any great extent, such as would warrant the planting of a much larger area in those sections where climate and soil have been found suitable for hop cultivation.

Even additional taxation of beer has not materially restricted consumption in the past and is not likely to in the future. Duties on hops would affect their value more than taxes on beer, yet the world's supply of hops must in the long run govern prices. In a year of short crops in the United States, a tariff of fifteen cents per pound on imported hops would be of more benefit to domestic producers than a duty of eight cents; in a season of domestic overproduction, the higher rate would not much influence the price of domestic hops, except possibly the fancy brands. In either case, the higher duty would not affect the price of beer, and therefore a moderately stiff tariff on imported hops is a good thing for American growers. But as "the foreigner pays the tariff tax," it would be bad for hop growers outside of England if the British Parliament should impose a high duty on hops imported into Great Britain, which is the market for the world's surplus of hops.

It is fortunate that the hop area throughout the world is limited, because, with an increased area available, the temptation would be such, in seasons of high prices, as to induce farmers to increase their acreage so as to thoroughly demoralize the market and depress prices to a point that would cause a loss to all growers. Such a condition has been experienced already more than once. Then, again, the failure of the hop crop in Europe has caused a heavy shortage in supply, with an extra demand for hops of American growth, for which abnormal and unhealthy prices have been paid,—unhealthy because they gave a temporary fictitious value to a staple crop, values which growers cannot with any show of reason or certainty expect to realize once in ten or twenty years. Yet the very fact that such a price as $1 per pound has been paid for hops

16

grown in this country, has stimulated farmers to largely increase their area and even to plant hops in locations that are not naturally adapted to their successful growth. The result, of course, has always been an oversupply with a heavy, dull, dragging market during several years, when dealers secured the crops at their own prices, which were not enough to pay the farmer for the actual cost of production. These periods of overproduction were followed by the destruction of plantations, with a consequent loss of time and money, till the market readjusted itself and became more settled. Then, again, the temptation arises to increase the production.

The wide fluctuations in the price of hops in the past are therefore easily accounted for. The most sensational was the advance to $1 and over per pound of the American crop in the fall of 1882, and a decline to 5c per pound three years later. Prices have since covered a wide range every season, though not to so marked an extent as in the instance cited. The crop of 1893 was comparatively short as a whole, following only medium crops for two or three years previous. This led to an increased acreage; with favorable weather the next two crops were the largest on record, and prices of the 1895 crop fell fully as low as ten years earlier. Growers seemed to have forgotten the lesson of the early 80's and made the same mistake a decade later. In this, however, the hop planter is no different from other people, for humanity has continued to make the same mistake generation after generation.

"The hop industry is a gamble," has therefore come to be an axiom. Yet with all its uncertainties this saying is not exactly true. Men who most perfectly understand the crop and most prudently allow for its uncertainties, have kept right along raising hops year after year, aiming at marketing about an even quantity of nice goods each season, and have found the industry rather more profitable in the end than any other crop grown in their neighborhood. It is fair to say that such men are a minority, and that the majority of American hop planters during the past forty years have quit hop growing poorer than when they began.

Much can be done to reduce the artificial uncertainties in the hop industry, also to mitigate the natural causes of variation. One object of this book is to set forth how this can be done, and thus to place the whole hop industry on a surer basis.

FIG. 4. A HOP HARVEST IN NEW YORK STATE.

This photo was taken in one of the large hop yards of Waterville, Oneida Co. The pole system is still much used in central New York, whereas the wire trellis seems to be preferred on the Pacific coast.

USES OF THE HOP

The manufacture of beer and ale consumes probably 95 per cent. or more of the world's production of hops. The oil from hops (that is, from the strobiles) is used for medicinal purposes. A decoction of hops is used in medicine for their tonic effect. Hops also have a sedative action, and are prescribed for derangements of the digestive organs attended by nervous irritability. The hop extract or lupulin kept in drug stores is preferred to the decoction for medicinal use. For hot applications to the body, nothing will retain heat or is more convenient for this purpose than a bag or compress of hops. For a variety of purposes, in household medicine, the hop is indispensable and widely used, as well as for yeast. Hops are prepared with a strong decoction of hops, oatmeal and water, and make an excellent remedy for ulcers, which should first be fomented with the decoction. A hop bath to relieve pain has also been recommended by physicians for certain painful internal diseases, made by boiling two pounds of hops in two gallons of water for half an hour, then strain and press and add the fluid to about thirty gallons of hot water. A pillow of hops induces sleep. Hop tea is said to be good for the blood and for fever.

The hop root contains much starchy matter and considerable tannin, but has never been utilized for these substances. The root has been used as a substitute for sarsaparilla. The tender shoots, taken when they just appear above ground, are cooked and eaten like asparagus or greens, making a dainty bitter relish, if the soil has been worked up so that the shoot is white for a foot or more. Hop buds are also used as a salad. The stem of the hop plant contains a vegetable wax and sap from which can be made a durable reddish brown. Its ash is used in the manufacture of Bohemian glass, and the vine also makes an excellent pulp for paper. From its fiber, ropes and coarse textile fabrics of considerable strength have been made. To make hop cloth the stalks are cut, done up in bundles and steeped like hemp, then dried in the sun, and beaten with mallets to loosen the fibers, which are afterward carded and woven in the usual way. Excellent paper and cardboard can be made from hop vines or roots, or from spent hops, and there are various patents and processes for such products. The vine being hollow, it is often used by boys for smoking purposes or as stems for pipes.

FIG. 5. A SOUTHERN OREGON HOP YARD, READY TO PICK.

Hop vines are usually burned after the crop is gathered, but if pressed into stacks or pits while still green they make an ensilage that is good feed for cattle. In France, the fresh hop leaves are also saved and fed with other forage to cattle. Valuable experience on this point is afforded by T. M. Hopkins of Worcester, England, who writes: "In October, I made two stacks of hop vines 16 by 16 feet and 18 feet high. After letting it ferment freely it was pressed down with a screw press and the next day was filled up again, and when sufficiently fermented, again pressed down, this process being repeated all through the hop picking. By March I had used nearly the whole of it, and calculate it saved me some 80 tons of hay. My horses have had nothing else for two months, excepting their usual allowance of corn, and I have never had them looking better. I have also had 100 head of cattle,—stores, cows and calves,—feeding on it, and they do well, the flow of milk being increased. Dr. Voelcker has analyzed it and says it contains plenty of good material, is decidedly rich in nitrogen, nor is the amount of organic acid excessive or likely to harm cattle. Another chemist says it contains more flesh-forming matter and less indigestible fiber than hay. Planters should leave off selling hops at a loss, but let the plant run wild, and they may every season cut two or three immense crops of material that will make silage of unexceptionable quality."

CHAPTER III.
CHARACTERISTICS OF THE PLANT

THERE is but one species of hop, *Humulus lupulus* though there are several varieties. The hop plant is naturally dioecious; that is, the male (staminate) and female (pistillate) flowers occur on different plants. Occasionally in a hop yard will be found what is called a hermaphrodite or bastard hop, with staminate and pistillate flowers on the same vine; the hop is not over half size, deformed, and is seldom gathered. Sometimes there will be not over one bastard vine to the acre, then there will be a dozen in half an acre. The bastard seems to be dwarfed, for it will go only one-half to two-thirds up its support. This sport does not seem to be permanent, for it seldom occurs twice in the same place. This freak usually occurs near a male vine, and there the female vine is so overcharged with pollen that it partakes of or is trying to represent the two genders in the same vine.

The hop is perennial; once started, from either root, cuttings, or from seed, the vine comes up anew from the same root year after year. The hop root is of a tough, leathery, spongy, porous nature. The hop has two distinct roots, a lower and an upper root, or runners. The lower roots have no eyes and propagation cannot take place from them, their office being to sustain the plant. The upper or surface roots have eyes or joints every four to six inches, their office being mainly for propagation. These are cut into pieces of two joints about six or eight inches long, for planting.

The root of the female plant is the lighter colored of the two, and the buds or eyes are more blunt. The male root is of a darker or grayish color, and the buds or eyes are more pointed and of a reddish or purple color. In America, a male root is planted for every 100 female hills; in England, one for every 200 to 300 hills; in Europe, the male plant is not countenanced.

The eyes are on opposite sides of every joint of the root. Each joint can throw out from six to a dozen buds. On a small root the center bud starts first, while on a large root, half a dozen buds start at the same time, each striving for the mastery.

FIG. 6. HOP YARD TWO MILES LONG AT HORSTVILLE, CAL.

Usually the vine that bears the hops comes out directly above the crown, but a surface root may run under the ground one foot or two feet, and then come out and run up its support and bear hops. These vines, when young and green and fresh, can be layered, covered with moist earth, and they will grow into roots with joints and eyes. The great objection to layered roots is that the joints will be too long and not as desirable for planting as the runners that come out naturally with shorter joints. From four to twenty or more vines will come out of every vigorous hill, and after selecting the desired number of the best for tying, the rest are destroyed and kept down by cutting or covering with earth.

FIG. 7. BRANCH OF MALE (Staminate) HOP VINES.
Reduced in size, and showing at the lower left-hand corner a single flower of the natural size.

The vines are put on the strings or poles when about two feet long. vines have to be put around the poles and tied with a string, but when strings are used to guide them to the wire trellis, it is only necessary to twine the vine around the string a couple of times, when by its innumerable little hooks on its six sides it will require no more attention unless shaken off by some violent motion, or by a continuous wind for several days. Should a three days' wind blow the vine off from the perpendicular string to the extent of a foot or eighteen inches, if the wind goes down at night, every vine will be found clinging to the string in the morning, having caught on again by their spiral or revolving motion.

Vines have to be put around horizontal strings or wires by hand. When left to their own inclination, they will grow upward until they become so long and heavy they will fall down and have to be replaced on their support. The end of the vine during the growing season,—say from one to two feet,—is very tender in the morning, and is easily broken; in the afternoon it will stand more careless handling without breaking. If the end of a vine is broken off in the growing season, the next joint will throw out two vines and soon catch up with the original vine and bear just as many hops, but the arms from the second joint are best.

FIG. 8. FEMALE VINE, SHOWING FLOWERS.

The vine when climbing a perpendicular support always winds with the sun, from left to right, or with the hands of a clock; other kinds of vines mostly climb in the opposite direction, which is, perhaps, why the patent office years ago granted a patent to a man who claimed to have "invented" the hop's habit of winding from left to right. The hop vine is hollow, six-sided, and has six rows of small, sharp hooks. These hooks are especially sharp on the tendrils, enabling the latter to cling fast, so that the plant can climb rapidly.

The hop vine has two motions. The first motion is a twist of the vine from right to left, the reverse of the sun and clock hands. The second motion is a spiral or revolving motion, with the sun, which winds the vine around its support. Whenever a vine is changed from a perpendicular to a horizontal position, the twist in the vine changes, or reverses, and twists with the sun. The instinct of a hop vine seems to be to follow a perpendicular position, and it cannot be made to follow an angle less than forty-five degrees without artificial means.

FIG. 9. BRANCH OF FEMALE HOPS.

A hop vine is one of the most ambitious of nature's climbers. It will go to the top of its support, if 20 to 30 feet high, and the hops will be on the extreme end, while none will be within 10 to 15 feet of the ground. When a shorter support is used, the arms will hang nearly to the ground, loaded with hops. The vine expands in diameter when four to six feet from the ground, where it will be nearly double the size it is one foot from the ground.

The leaves of the hop vine are irregular in size and conformation; the larger are usually three to five lobed, the smaller heart-shaped. There is no fixed characteristic difference in foliage between varieties.

The flowers are very numerous on both male and female plants. The male flowers are in loose clusters, as shown at Fig. 7, of a yellowish green color, with a five-pointed calyx and five stamens. The male plant produces the pollen, which, carried by the wind or by bees or insects to the pistil of the female flower, fertilizes the latter so that it can produce seed. Unless there are a sufficient number of male plants (say one to every 100 female roots), no seed will be produced. The male vine bears no hops. The male flower in bloom produces a very fine flour called pollen. This pollen can be seen in the morning, when the light is just right, by vibrating the male vine and looking toward the sun. It looks like the dust in the rays of the sun when passing through a knot hole.

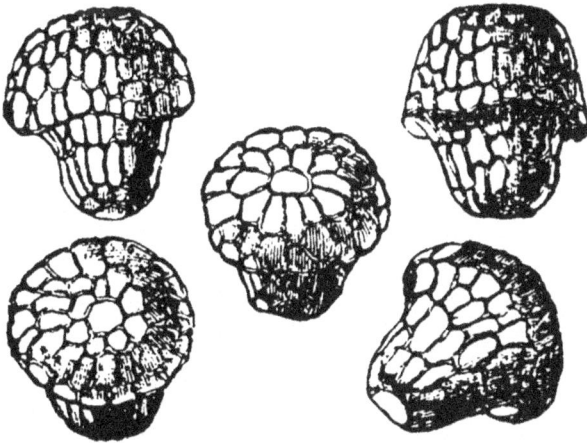

FIG. 10. GRAINS OF LUPULIN.
Highly magnified.

The female plants, only, produce hops. The single cone of hops is botanically called a strobile, and consists of a series of scales or bracts and their fruit. The female flowers are borne at the base of these scales, which are arranged in close clusters on a short stem. When in blossom, the young hop will be found to be a collection of very simple flowers, each consisting of a single pistil surrounded by a sort of membranous covering, and one of these is placed at the base of a small scale, which, as the hop ripens, increases very much in size, and collectively becomes the most conspicuous part of the cluster of fruit or hops, Fig. 9. The fruit, botanically speaking, is the ripened pistil, which is a small nut that incloses a single seed. Upon the inner side of the scales, and around the fruit, are found numerous yellow grains which are peculiar glands; and, though they are produced only in the pistillate plant, they are often incorrectly called the pollen. These grains are called lupulin, and sometimes "lupulinic glands" and "flour of the hop."

FIG. 11. FEMALE CLUSTER, NEWLY SET.
b, Hop cluster.

The female flowers are in the form of a catkin, having each pair of flowers supported by a bract, which is ovate-acute, tubular at base. Sepal solitary, obtuse, smaller than the bract, and enfolding the ovary. Ovary roundish, compressed; stigmas (the terminals of the pistils) two, long subulate, downy. The bracts enlarge into a persistent catkin (hop), each bract enclosing a nut enveloped in its permanent bractlet, and several grains of yellow lupulin.

FIG. 12. SINGLE FEMALE FLOWER.
a, Pistils; *b*, scales; *c*, single seed with its scales.

The leaves on the strobile or hop point outward and look rough until the pistil has been properly fertilized with pollen from the male plant, when they close down and become smooth, four-sided cones. If no staminate or male hops are in the yard or vicinity, the points of the leaves will point outward, giving the hop a rough and imperfect appearance.

SEEDLESS VERSUS SEED HOPS

Whether male plants are an advantage or a disadvantage to the hops of commerce, has long been disputed. It is not necessary that the female hop should make seed in order to maintain the strength and vigor of the plant, although the contrary opinion is much held. Indeed, continuous ripening of seed is one of the most serious drains on the vitality of hop roots. Plump seeds comprise about 10 per cent. of the weight of the cured hops. Hence, from the standpoint of the seller, they are not to be discarded without good reason. This matter has been laboriously investigated by German scientists, who seem to be agreed that fertilization lessens the quantity of lupulin, and injures its quality by making it less oily and less aromatic. "The function which the plant would use in ripening seed seems to be employed in forming lupulin more abundantly, and in making the hop fine, and imparting to it the peculiarly rich aroma so much desired by certain brewers." So true is this held to be that male plants are not permitted in Spält under heavy legal penalty, and a Belgian commissioner appointed to specially study this whole matter concludes: "Banish strictly all male plants from your hop yard."

FIG. 13. CROSS-SECTIONAL LONGITUDIANL VIEW OF FEMALE HOP.

European authorities also maintain that "fertilization increases the number and appearance of the cones; they become coarser, looser, and longer, and the bracts are longer, more brittle, and fall off more easily." They estimate that 116 pounds of seed hops are required to get an equivalent effect in the beer of 100 pounds of seedless hops.

Prof. Cheshire, who has made a special study at the Kew gardens, London, of the relations between insects and flowering plants, also agrees with European authorities, and says: "The scientific evidence is all on one side— that for the production of the largest percentage of lupulin, fertilization should be prevented by suppressing the male plant. As a set-off against this, however, fertilization (which directs the energies of the plant to maturing its seed) absorbs into the seed a very large part of the store of nutritive material at the disposal of the plant, thus increasing the actual weight of the crop of hops by about 10 per cent. This increase in weight is accompanied by a considerable percentage decrease in lupulin and aroma, the very matters for which the hop is grown. The question is, therefore, to be settled entirely on commercial lines, whether quality or bulk will bring the larger returns."

Editor Ironmonger of the English *Hop Grower* also concludes his inquiry thus: "If the hop grower wants exquisite aroma and fine condition, he must exclude the male plant and stop fertilization. If he wants weight with the sacrifice of some quality, let him encourage the males and gather his well-seeded, heavy crop."

On the other hand, the subject is claimed to be a matter of taste. Those who like the German beers made from seedless hops do not like beer made from English or American seed hops, and vice versa. But the increase in the sale of American and English beer has outstripped that of their German competitor, indicating that the demand is not so fastidious about seed hops as some people think. Further testimony on this point is afforded by the fact that even Spält hops, which command the highest price and come from a region where the utmost effort is made to exterminate the male plant, contain a goodly proportion of seeds. If it is proposed to compete with German hops in the German market, or to displace German hops that are yearly sold in moderate lots at fancy prices to both English and American brewers, then the male plant must be extirpated, and every effort made to closely imitate the peculiarities of the German marks. This special market is to be got by catering to its whims, not by opposing them.

FIG. 14. VARIOUS SHAPES OF HOPS.

Aside from this special and limited demand for German seedless hops, it is evident that the bulk of the trade does not particularly care about hops being seedless. It is significant, also, that the objection to seed hops is mainly heard when prices are so very high that this point is raised really as an excuse for a lower quotation, or for an allowance because of the weight of the seeds, which constitute about 10 per cent. of the total weights. The consensus of opinion among American expert dealers and growers, whose views have been carefully collaborated by the author, agrees that imperfect fecundation is a frequent cause of light weight hops of inferior quality. Especially important testimony on this point comes from A. J. Wolcott, an experienced grower in Polk county, Oregon:

FIG. 15. KENTISH HOPS.

"This complaint of the Germans of seeds in American hops was first heard in 1882 when hops were so high, and caused some growers on this coast to grub out and destroy all their male vines. The result was that their hops did not mature well. They were large, green, light, feathery things, with neither color nor strength, and dealers would not handle them. I have seen this experiment tried in southern Oregon with the same result. I planted a yard myself once without being able to get male roots, and my hops were poor, lean things, until I obtained the male plants and got them to growing vigorously, when my hops became of good color when ripe, with plenty of strength, and I heard no more complaints of poorly matured or lean hops. I am now fully convinced that hops, like many other plants, require fertilizing from the bloom, and as none but the male hops bears any pollen, it is necessary to have a sufficient number of these in a hop yard so that the female flowers of each vine may be fertilized. And brewers, if they expect a good solid, bright-colored, well-matured hop, well filled with lupulin, must expect also to see the hop well filled with good, large purple seed. If they do not wish seed, they cannot expect lupulin. Germany may produce good hops without seed, but it cannot be done here, at least such has been my observation and experience. Therefore, my advice is, to let the male hop alone and if in a season of high prices a few brewers complain of extra weight in the seed, pay no attention, but go ahead."

VARIETIES OF HOPS

Here there is "confusion worse confounded." Plants raised from seed are new varieties; only root cuttings propagate the same varieties. Many varieties have been produced and as it is difficult to distinguish between their roots, the "sets" have been more or less mixed. There has been an astonishing lack of care to preserve the purity of the best varieties, and as the same common name is sometimes applied to roots of different varieties when grown in other sections, there has come about an almost hopeless confusion of varieties. This is especially true of America, and to a less extent in England, while the best continental growers jealously guard against such confusion and insist upon sets true to their pet varieties. Such care accounts in part for the peculiar merits of certain brands of English and European hops.

Aside from the exceptions just stated, it is a curious fact that there have been no real efforts to breed improved varieties of hops. There is an old saying that there are already too many kinds, but there are not too many varieties of No. 1 hops. And, too, these best varieties are probably chance seedlings, instead of being bred from parents selected for some known good qualities. It is also doubtless true in a measure that the constant propagation by cuttings, having been carried on for many generations, has caused some loss of vigor and constitution, which may account for hop yards being more easily affected by fungus and insect pests now than half a century ago. Nature's law of reproduction is by the union of sexes and she also opposes in-and-in breeding. If the experiment stations in New York or on the Pacific coast would take up this matter, there is no doubt that in a few years much could be done to improve the hop crop by careful selection and hybridization. A correspondent in the English *Hop Grower* of February 19, 1895, suggests the following method:

FIG. 16. KENTISH CLUSTER HOPS.
At Watsonville, Cal., from a photograph taken Sept. 10; yielded 2000 lbs. Of dried hops to the acres, as the average for 19 acres. A general view of this yard, trained on short poles, is shown on another page.

"As the male is generally supposed to influence the constitution, and our aim is to produce a vigorous, disease-resisting hop of good quality, let us take, by way of example, a Fuggle for the father (though it has been noticed that male Fuggles are scarce in East Kent plantations), and for the mother we will take a Brambling or Petham Golding, as being of the best quality. In the autumn, select strong, healthy hills from which to take the cuttings, and plant the male Fuggle cuts and the female Brambling cuts a short distance apart, the male preferably on the southwest side in a sheltered spot, as far as possible away from any hop garden, to guard against the female being fertilized by any but the desired pollen. The first year we shall probably get very few or no hops; the next season perhaps it will be advisable to cut the Fuggle early, and the Brambling late, in order to cause them to flower as nearly as possible at the same time, one being an early, the other a late hop. In the autumn the crop must be protected from birds and be allowed to get thoroughly ripe; the hops must then be picked and sun-dried; afterwards the seeds must be sorted out and kept till the spring, when they can either be sown in pots or in a bed properly prepared. In the autumn the seedlings must be dug up and planted out on good ground about two feet apart, and carefully cultivated for two or three years, till it can be seen which plants answer our expectations and are worth saving; these can afterwards be readily reproduced from cuttings."

In England.—As to varieties in England, Whitehead wrote in 1893 as follows, and careful inquiry by the author, of the best experts among English growers, shows that this is equally true to-day:

"The fashion as to varieties changes, in accordance with the circumstances of the demand. Until the last year or so hops of the finest quality were required by the brewers. Land which produced these was at a premium. The East and Mid Kent and Farnham planters were in the ascendant and planted the best varieties, as Bramblings, and others of Golding character. Producers of more common hops, in the weald of Kent, Sussex, and elsewhere, were disposed to consider their occupation gone, and made some efforts to improve their quality. But now this has changed for the nonce. Fine-flavored hops, full of aroma, seem just now to be required only for pale and export ales, and for the comparatively small quantity of stock beer now brewed. For beer for quick draught common hops, it is said, are good enough. There has been a large demand for these of late, and they have made prices relatively higher than those of the finer sorts. Varieties of common hops have therefore been extensively planted, even in districts producing hops of fine quality, and among them the Fuggle's Golding, as cropping heavily, has been largely selected. Many planters, however, refuse to make any alteration in this respect, as they say that there will be a reaction when the market is crowded with common hops.

"In East Kent the prevailing varieties are Goldings of several kinds, Bramblings, Cobb's Early Goldings, Petham Goldings, Canterbury, and Old Goldings. Bramblings and other Goldings are still generally grown on the best land; Whitebine Grapes and Grapes on that of not so good a quality. In Mid Kent, Goldings, Bramblings, Grapes, and Jones are principally cultivated. Fuggle's Goldings are now being planted rather extensively.

"The Golding is undoubtedly the best English hop, having unsurpassed aroma and brewing value. The Golding is a sub-variety of the Canterbury hop, which was raised by a Mr. Golding of Kent, about 1800, who observed in his grounds a plant of extraordinary quality and productiveness and marked it and propagated from it, furnishing his neighbors with cuttings. This variety has small compact cones, shaped somewhat like a filbert, of a light golden color when ripe. The cones do not cluster together, but grow in bunches of two or three cones. Bramblings are Goldings of slightly different shape, coming earlier to pick, having valuable Golding attributes. White's Early Golding is the earliest hop with Golding characteristics, but it is rather delicate, and a shy bearer.

"The Grape and Whitebine Grape are very useful, handy sorts, having large cones that grow to a great size in some soils, and hanging in clusters like grapes. There are other kinds of Grapes, as the Farnham Whitebine, full of quality and a very good bearer. Cooper's White is a rather early variety. Mayfield Grape is a hardy, useful prolific kind. Buss's Golding and Fuggle's Golding have not many Golding qualities. They are rather coarse, coming to pick later than Goldings, but they are good cropping sorts, especially the Fuggle's Golding, and are not as a rule so disposed to blight and mold as others. The Jones is a very useful hop, yielding well on some soils; it has large cones, and when grown on good land has much quality. There are very early and common varieties, as Prolifics, Meophams, and others, which yield large crops of inferior quality, and are not much in favor with brewers when other kinds are available at reasonable rates.

"The Mathon is peculiar to Worcestershire and Herefordshire, and approaches nearly in flavor to the East Kent Golding. In Sussex and the weald of Kent, the Colegate is grown, but not nearly so extensively as twenty-five years ago, and many planters are eliminating it altogether and planting Fuggle's, Hobb's, Henham's, and Buss's Golding. It comes to pick latest of all hops. It is a very hardy but backward hop, and will grow on any soil; it runs much to vine and requires as long poles as Goldings. The hops are generally very small, when quite ripe before they are picked; they have a rich, thick appearance when dried, but the smell and flavor are not good, and some brewers object to them.

"Hops of a Golding type are cultivated on the best soils in Hampshire and Surrey, while Grapes, as the ordinary Grape, and William's Whitebine Grape, the Grape Greenbine, Henham's and Fuggle's have been planted on the poorer soil. There has been a disposition of late in Herefordshire and Worcestershire to plant hops of Golding character, and to improve the quality generally of the growths of these counties, which finds much and increasing favor among brewers. At the same time early varieties, as Meophams and Prolifics, have been put into some extent, and Fuggle's, which are coming into favor."

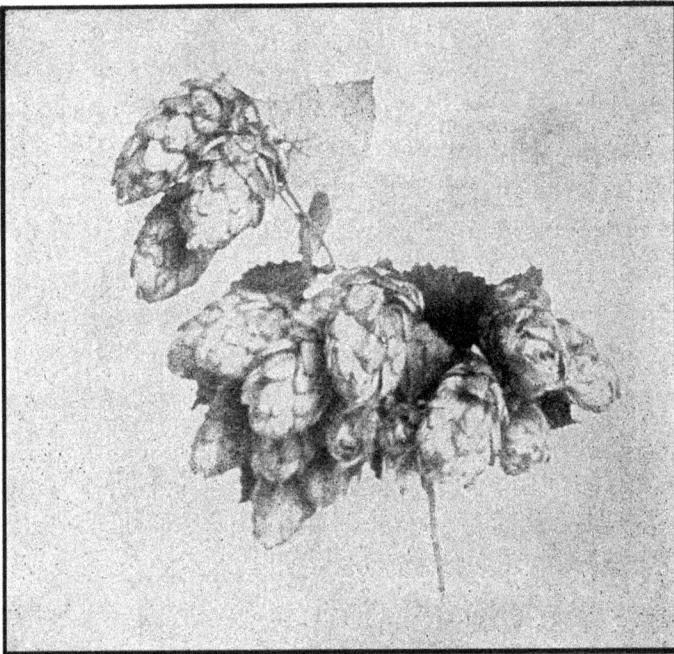

FIG. 17. FUGGLES, KENT.

37

In the United States, the number of so-called varieties is much smaller than abroad. In New York state, English Cluster is still grown extensively, being strong, a vigorous climber, and bearing rich, golden hops when well handled. The Grape is a very rich hop, but not so hardy nor so good a climber as English Cluster, which has largely supplanted it. The Grape vine and fruit are of only medium size, the hops have a mild flavor and part very easily from the stems. Pompey has large, rank-growing, rough vines, dark green foliage, large, squarish and strong-flavored fruit, sometimes three and even four inches in length, is hard to pick, and is no longer planted by progressive growers. About a week earlier in ripening is Humphrey's Seedling, originated by chance in Wisconsin, which is being grown quite extensively. It is a good grower, but sometimes yields lighter weight hops than Cluster, lice are very partial to it and the yield is sometimes cut by a hot, dry spell coming just as the hops are in the burr. One week earlier than Humphrey is the main point of Palmer Seedling, but it is such a shy bearer, though of fine hops, as not to be much planted.

The Canada hop or Canada Red, so-called because the roots come from Canada, is known by its red vines, fruit rather below medium size; the strobile is firm, of a golden color, and mild, agreeable flavor. It is perhaps the hardiest of all hops, and seldom winterkills in New York state, when other kinds may be ruined. It is a fair bearer under indifferent culture, and a good bearer under good culture. The hops are leafy and rather difficult to pick clean, which probably accounts for the dispute as to the flavor and quality of the Canada hop. "It is of rank flavor and disliked by brewers and dealers" when moldy, unripe or overripe, or when mixed with leaves, etc.; but picked clean in its prime and properly cured, the true Canada hop is of fine flavor and color, though perhaps not as good as English Cluster. The popularity of Canada is due mainly to the fact that it ripens nearly a week later than Cluster and can stand on the vines fully a week after the date that Cluster must be picked. The roots are also cheaper.

A false Canada, or roots of an inferior quality, has been spuriously sold. It is such hops that are usually "so rank in flavor and disliked by the trade" as to be a commercial failure. The term "false Canadian hops" is not recognized in Canada. It is a fact, however, that Canadian hops are so disliked in England they cannot be sold there. John A. Morton says Canada produces mainly three kinds: "A hop that grades very similar to the best growth in Franklin county,

New York, another akin to English Cluster, but with a slight Bavarian flavor, and a third variety very similar to Pacific Coast hops."

As to *California* varieties, Flint writes for this book: "There are only two varieties of hops cultivated here to any great extent. The leading variety is called the large gray American hop. The hop is large and compact on the stems. We are so well pleased with it in every respect, except that in some localities it does not give as fine straw color as we would like, that we are not looking for a better one. Another variety is called the 'San Jose hop,' but the growers do not plant it if they know it. It comes a little sooner in the spring and outgrows the other kinds for a while, has more and larger leaves, but the hops are more scattered on the arms and do not produce as much per acre. The vine is redder or darker.

"There have been a very few importations of some of the leading varieties of the English hop, but not enough to make a commercial account of them. I learn that they do not produce equal to our kind, and I think they will be soon discarded.

"In the Sacramento valley we begin to pick the American hops about August 23 to 28; nearer to the coast, about ten days later. When hops are ready to pick they will not stand more than three to four weeks without turning rusty, but this depends a good deal on the weather; a hot north wind will burn them like a furnace in one day, on the sides of the yard most exposed. On the other hand, if the weather remains cool, and we have cool south winds and nights with a good deal of dew, the hops will remain on the vines much longer without turning. Most all the hops on the Pacific coast came from my yard and I brought the stock originally from Vermont in 1855."

In *Oregon,* the English Cluster, or White Root hop, produces 1,500 to 2,000 pounds per acre. It usually ripens between September 1 and 10, and will stand 15 to 18 days, when the hops begin to turn red or overripe and to scatter. It is the heaviest yielder and brewers prefer its quality, but its roots are especially subject to black knot. Canadian, or red root hops, or as it is sometimes called in California the Bavarian root, yields less in Oregon, averaging 1,000 to 1,500 pounds per acre, but the crop is surer than English Cluster, ripens seven to ten days later and will stand three weeks, but the strobile is small and the foliage thick, so it costs more to pick and sells no higher. Early Fuggles are passing away. They produce a small hop, almost seedless and have a peculiar flavor. They ripen from ten days to two weeks earlier than the English Cluster and even when ripe they have a dull, greenish color that detracts from the selling qualities.

FIG. 18. BATES'S BREWERS, KENT.

The rotative dates of maturity are important, as a yard can be more conveniently harvested if parts of it ripen in rotation. Palmer ripens first; Humphrey's follows a week later. English Cluster matures five to ten days afterward and true Canada Red about a week later still. E. Meeker thinks that a model yard in respect to varieties for the Pacific coast is in the proportion of one acre of Humphrey to two acres of Cluster and one acre of Canada Red.

European Varieties—In Germany, the merits of hops are governed more by the sections in which they are grown than by the variety. Beckenhaupt says that this is true to so great an extent that only three distinct varieties of hops are grown in Germany—the early Schwetzinger, the medium Rothermburger and the late Spälter variety. Beckenhaupt has collected nearly thirty distinct varieties of hops from various parts of Europe, and believes there is more merit in the best varieties than is recognized by the trade. Most of the hop yards in Germany are planted to roots that came directly or indirectly from Spält, and the hops there are said to have been derived from Saatz hops. Evidently the confusion of varieties is quite as bad on the continent as elsewhere.

Over 100 so-called varieties of hops have been described in Germany, but in the Nuremberg market continental hops are thus classified, and the varieties grown in the respective localities are much alike, if not identical: I, Hops of the towns of Saatz and of Spält, and the nearest situated principal villages; II, adjoining domain of Spält, Kind, and Saatz lands; III, Wolzach, Au, and smaller sites of the Spält land; IV, Hallertau, Auscha red-land, Styria, and principal portions of Wurtemberg and Baden; V, finest mountain hops, Aisch-ground, finest Polish, Alsatian and Burgundian hops; VI, Common Middle and Upper Franconian hops, Wurtembergian, Baden, Polish, Alsatian and Burgundian, and finest Galician hops; VII, Upper Austrian, Auscha Greenland, Lothingian, and Kannenberg land; VIII, Brunswick, Altmark, and the remaining parts of northern Germany; IX, northern France, Belgium and Holland; X, Russia and the rest of Europe.

WHAT CONSTITUTES QUALITY IN HOPS?

This question can be answered in various ways, according to the peculiar desires of the buyer. What Brewer A might consider specially adapted to his need, might be considered as No. 2's by Brewer B, whose product was quite different from A's. Dealers and brewers, also growers who know their business, are united upon certain characteristics as essential in any lot of hops in order that it may command the top of the market:

1. The hops must be picked clean, and be free from leaves or weeds, with no pieces of vine, string, or dirt, and free from discolored or moldy hops.

2. The hops must have a rich, golden color, not over-bleached nor yet too green, properly cured but not baked, with a bright attractive appearance and a rich aroma.

FIG. 19. REAR VIEW OF 12 KILNS, PLEASANTON, CAL., WITH GROUP OF CAMP-FOLLOWERS IN THE FOREGROUND.

3. They must be honestly and solidly packed in bales weighing about 185 pounds, and in baling great care should be observed not to break the hops, nor to false-pack them. The term "false-pack" is applied to a bale of hops that has two or more colors of hops, pressed in layers. It is caused by filling up a portion of the bale with hops of a certain color and then filling the balance with hops of another color. When a sample is drawn from a bale of this kind, one portion of the sample will show hops probably of a greenish character, and the other portion will show up a yellow or red. It is essential that the bale be even in color and that a sample taken from any part fully represent the whole.

Viewed from the standpoint of brewers as brought out at a recent conference of English hop growers, it is necessary to consider three things in judging quality of hops: First, the preservative power, depending upon the particular class of resin called oleo resin; second, the amount of bitter that is yielded, and third, the flavor or aroma. Generally speaking, the higher the percentage of oleo resins, which are now capable of exact determination, the greater the value of hops for brewing purposes. Discussing these features, Mr. C. S. Meacham, a brewer of Maidstone, Eng., among other things said:

"Continental or American hops, growing a high percentage of oily resins, are generally so intensely bitter that it is this which determines the maximum quantity which a brewer should use in his beers in conjunction with his milder British hops. It must be borne in mind, however, that the amount of the bitter is to some extent counterbalanced by the lesser quantity of hops which need be used where the percentage of preservative resins is high. This statement might be construed by some as a reason for not striving after a high resin percentage, but I would remind such that it is not bulk of hops but preservative resin which the brewer wishes to purchase, and for which he will be prepared to pay more money.

It is on this point of resin that the battle between English and foreign hops is to be fought. In softness and clearness of bitter and delicacy of flavor, English growers have little to fear; but from the point of view of resin percentage, continental and American competitors so far have the advantage. To reduce the difference between British and foreign hops to a small point, the end may be attained by allowing British hops to fully mature before picking, by greater care and delicacy in handling, and by greater care and supervision in drying. Careful investigation shows that a great loss of resin is due to careless handling, in some cases amounting to 20 or 30 per cent. of the total. In German hops scarcely a particle of the resin is lost. The one question now before the hop growers of England is how to produce the largest amount of resin preservative and then to save as much of it as possible. If the English curing can be improved so as to give a return of 18 per cent. resins, instead of about 15 per cent., as at present, the difference between English and foreign grown would be infinitesimal."

FIG. 20. CLIMBING TENDRIL OF HOP VINE.
Magnified 100 times, showing the prickly points that give the vine
such power to cling.

BREWERS' VIEWS IN BUYING HOPS.

[Quoted from The Theory and Practice of the Preparation of Malt, etc.]

"The brewer, in determining the value of hops, is forced to take into consideration certain external qualities, far more so than in barley, for he can reach a conclusion as to suitableness of the hop for the fabrication of beer from external appearances alone. We will here give the good as well as the bad qualities of hops.

"1. The cones of the hop should not be too large; the carpels should not be thick and leathery, but tender, and their ribs should be thin. The color of the cones should be yellowish green and not light green, red, or reddish brown. The peduncle should not be stripped of leaves and loose carpels should not be mixed with the hops in large quantities, but the cones should appear closed, with the carpels lying tightly above each other. Cones of a light green coloring and open are frequently proof of unripe hops, which contain less flour and have a weaker aromatic smell. A light red coloring and a very shiny surface of the carpels is an indication of the hops having been allowed to become overripe. The consequence of overripeness is a loss of the valuable flour, yet this is not so injurious as when the component parts of the hops have suffered injury from having been heated during drying, and the hops have acquired a dull brown color in consequence. This appearance is called 'ground red' (*bodenroth*). The hops have a similar appearance when they have been baled too damp and have become heated in the hop bale, when they largely lose their agreeable aroma and very frequently become entirely useless. If the hops have been dried too much, or have been frequently repacked, for whatever reason, the carpels become detached from the peduncle, the cones appear to be torn, and they have lost some of their flour. If the hops have been dried by artificial heat, at too high a temperature, the flour assumes an orange color and the hops acquire an empyreumatic smell.

"2. When a few cones are torn to pieces, as large a quantity of hop flour as possible should be seen on the inner surface of the carpels. The richer the hop is in flour, which is the bearer of its most valuable component parts, the more valuable will it be, if it also possesses the other good qualities. The flour of fresh hops is a light yellow color. The fruit, situated on the base of the carpels, should be as small as possible; large granules which weigh heavy are an indication of a not very fine hop.

"3. A fine, strong aromatic odor should be perceptible when the cones are rubbed between the hands. Hops of poor quality, or raised under unfavorable conditions, possess a garlicky odor. Hops smelling moldy or musty, or which have suffered injury in drying, or in the hop bale, should not be used.

"4. The separate cones should stick together when the hop is pressed together in the hand—it should ball together and only slowly separate again; this is an indication of the hop being rich in resin. If it contains little resin it does not ball and feels dry. When marks are made upon the hand with separate cones, these marks should be sticky and of a yellowish color; unripe cones make light green marks.

"5. The taste of the hop should be pure and agreeably bitter.

"6. It should be free from the leaves of the vines, pieces of vine and other admixtures. The cone should not be covered with mold or the parasitic fungus-smut (*Fumago salicina*), which covers the leaves and cones with a sooty coating, and is very injurious to the hop plant. This fungus may destroy an entire hop harvest. Plant lice frequently make their appearance as forerunners of this disease, adhering in skins to the hop and contaminating it.

"7. They must not be too old. Old hops do not possess certain of those already-mentioned good qualities; they have lost considerably in value, as the volatile oil, as well as the hop resin, has deteriorated. Hop cones which have been stored for some length of time have a brownish color, the fruits are easily detached from the peduncle, the agreeable odor has changed into a disagreeable, cheesy (rank) odor, the flour has a reddish coloring and the hop has lost its stickiness. The examination of the hop flour by a good magnifying glass or the microscope is to be recommended as a means of distinguishing old hops from fresh ones. Even then, when the before-mentioned characteristics of old hops have in some manner been obliterated with fraudulent intent, a microscopic examination is still a sure guide. The separate glands of fresh hops, which form the hop flour, are full, glossy, and of a lemon color, have a smooth surface, and when pressed discharge the contents of the gland, showing a light yellow coloring. Glands of the old hops are shriveled, wrinkled, and the fluid discharged from them is of greater consistency and has a dark yellow to brownish color, and this color will show itself the more the older the hops are, and the smaller the quantity of hop balsam. In time the hops become poorer in oil, which has been partly oxidized and changed its color."

CHAPTER IV.
COMPOSITION OF THE HOP PLANT AND ITS FRUIT

THE essential characteristic of the hop plant is the lupulin in its flowers or strobiles. When highly magnified the grains of lupulin appear as in Fig. 10. When fresh, the lupulin is very resinous, adhesive and aromatic; and it is upon this that the peculiar odor, taste and other properties of the hop in a great measure depend. This being the case, the greater or less abundance of lupulin in a sample of hops is one guide in judging of their quality, and it will be seen that, in all processes of preparing them for market, care should be taken that this be not lost. The color of the lupulin is also an essential feature. A bright golden lupulin of a lemon shade is an indication of proper curing, while a discolored lupulin is an indication of improper curing or handling and is caused by the hops being either high dried or the reverse.

The odor of the hop or strobile is due to its essential oil and is powerful but agreeable; the taste is bitter, and besides lupulin, the hop contains an acid, an essential oil, an aromatic resin, wax and extractive matter. By pressure, hop heads yield a green, light, acrid oil, called oil of hops.

Yves attributes to lupulinic powder alone the active principle of hops. But Payen and Chevallier are of the opinion that the entire flower contains the same active principles which are found in the yellow dust. If this were not so, says Simmonds, the hops, which in transport lose a great quantity of this yellow powder, would have but a feeble effect in the manufacture of beer. The bracts certainly contain some lupulin and are therefore not altogether devoid of active principles. The tonic and narcotic properties of the hop are peculiar to it and occur in no other substance. Lupulin alone contains the following substances in varying proportions:

1. Water
2. Essential oil.
3. Acetate of ammonia.
4. Malate of lime.
5. Albumin.
6. Gum.
7. Malic acid.
8. Tannic acid.

9. A resin.   10. Bitter extract.
11. A fatty matter.   12. Chlorophyl.
13. Acetate of lime.   14. Nitrate and sulphate of potash.
15. Sub-carbonate of potash.   16. Carbonate and phosphate of lime.
17. Phosphate of magnesia.   18. Sulphur.
19. Oxide of iron.   20. Silica.

The well cured American or English hop of commerce varies in composition within the following limits, the figures showing per cents or pounds of each ingredient in 100 pounds of cured hops:

	Lupulin.	Water.	Seeds.
Highest	16	10	14
Lowest	7	7	8
Average	10	8	10

In German hops the seeds comprise from less than 1 per cent. to 8 per cent. of the weight, but their lupulin varies as above; of the cones, 75 per cent. are outer leaves or bracts and about 15 per cent. are stalks. Numerous analyses of hops have been made abroad, but few in the United States. They do not throw much light on the question of what constitutes quality in hops, for the constituents revealed by analysis have been found to be almost identical in two samples, one of which produced a good beer and the other a bad beer. Simmonds gives a compilation of analytical data about hops to which the student is referred, and further information will be found in books on brewing. More recently Aubry analyzed nine samples of continental and English hops, the results showing less variation than might be expected, as follows:

	Maximum.	Minimum.	Average.
Water	9.3	7.6	8.3
Alcohol extract	48.8	39.1	44.8
Ether extract	36.2	26.4	31.6
Water extract after alcoholic extract	16.3	11.2	14.0
Nitrogen, total	2.8	1.9	2.3
Nitrogen, soluble	0.8	0.6	0.7
Tannin	5.8	4.0	4.8
Resin	26.1	15.6	20.1
Water extract	27.0	21.1	24.2

FIG. 21. LARGEST HOP KILN IN THE WORLD UNDER ONE ROOF.
Meeker's large plant in the Puyallup valley, Washington, U.S.A.

The conclusion drawn from these results is that the tannic acid effects of hops are accomplished mostly after they have been deprived of their resin, and that but a very small proportion of the nitrogenous constituents of the hops being taken up by the water extract, they are of but little importance in the wort.

THE CHEMISTRY OF HOPS

is treated in further detail for this work by E. E. Ewell, of the Division of Chemistry, United States Department of Agriculture, as follows: "That our knowledge of the chemistry of hops is still deficient in many points is shown by this quotation from Moritz and Morris's 'Text Book of the Science of Brewing,' published in 1891:

"Although it must be granted that in recent years we have got to know something precise as to many of the constituents of the hop, yet its chemistry, like all botanical chemistry, is surrounded by difficulties in regard to the isolation and investigation of the various constituents, difficulties far exceeding those of the study of other materials used in brewing."

According to these authors, "Hops are added to the beer for the following reasons: (1) To give the beer the distinctive bitter flavor and aroma; (2) to precipitate certain nitrogenous constituents of the wort; (3) to clarify the wort, not only by the separation of the above constituents, but by the mechanical clarifying property of the hop leaves when agitated in the copper, and by the formation of a filter bed for the filtration of the wort in the hop back; (4) to preserve the beer by the antiseptic influence of some of their constituents; (5) to assist in the sterilization of the wort.

FIG. 22. A HOP HARVEST IN MADISON COUNTY, N.Y.

"The bitter flavor is imparted by some of the resin and the so-called hop acid (*hopfenbittersaure*); the aroma by the volatile oils; the precipitation of the nitrogenous matters by the tannic acid, and the antiseptic properties by certain of the resins. It is therefore essential that hops, to be of value, should contain these substances in due proportions.

"The percentage of tannic acid in hops is stated by communications from the Agricultural Laboratory of Vienna to range from 1.38 to 5.13 per cent., the average being between 3 and 3.5 per cent., and this, so far as we know, is the normal amount found in good hops. The percentage of volatile oil, shown by the analyses contained in the report from the Austrian laboratory named above, ranges from 0.15 to 0.48, the average being about 0.25 to 0.35. To these oils we owe the aroma and delicate flavor of the beer.

"The bittering principles of hops are still the subject of considerable divergence of opinion. According to Hayduck, the resins are the essential bittering principle, and as Hayduck's researches are the most recent and are characterized by completeness and definiteness, it is probable that his views are more worthy of credence than those of the older investigators. Amon these is Lermer, who claims to have separated a crystalline bitter acid from hops, to which he attributes their bittering properties. The acid is insoluble in water, but soluble in dilute alcohol, imparting to the solution an intensely bitter taste.

"Julich separated an intensely bitter substance from hops, which was easily soluble in water. Bungener attributes the bitter to a substance partially of an acid, partially of an aldehydic nature. The substance is insoluble in water, but easily soluble in alcohol, ether, etc. It is easily oxidized to valerianic acid, and Bungener attributes the presence of this acid in old hops to this cause."

FIG. 23. A PARTLY PICKED HOP YARD, CALIFORNIA.

Various figures are given for the percentages of the true bitter principle, but owing to the widely differing opinions in regard to the exact nature of the one or more bitter substances contained in hops, it is not thought wise to repeat these figures. Hayduck found at least three resinous bodies in hops. Data in regard to the percentages of these resins are not at hand, but Blyth publishes, in his treatise on foods, an analysis of lupulin by Dr. Yves, which shows 30 per cent. of resin.

Several analysts have devoted considerable time to the detection of an alkaloidal, or other constituent of hops, which will account for the narcotic or stupefying effect of beers, in the brewing of which large proportions of hops are used. Moritz and Morris, in their book already mentioned, state that while this view as first announced by Graham is reasonable, it is not at all improbable that the higher alcohols developed at the higher temperatures prevalent in the English practice of brewing are also important factors in producing a beer possessing a greater stupefying effect than the lager beers produced on the continent.

Griessmeyer reported an alkaloid to which he gave the name of lupulin. Griess and Harrow separated a base from beer which proved to be cholin. Griessmeyer denied the presence of cholin as such in hops, stating that it exists combined with other bodies as lecithin, a body of very complicated constitution.

Southby, in his work on practical brewing, states that by distilling hops in a current of steam he was able to obtain from 1.5 to 2 per cent. of volatile oil, quantities decidedly in excess of the others given above.

Moritz and Morris state that the chemistry of hops is still in such an imperfect state that physical characteristics, odor, color, etc., must for the present be relied upon in the judgment of this important brewers' raw material.

Moritz and Morris have prepared a summary of 26 analyses, which Wolff has published, of the ash of German hops. The average per cent. of mineral matter is 7.4, the maximum 15.3, and the minimum 5.3. Analyses of the ash showed the following percentage composition:

	Minimum. %	Maximum. %	Average. %
Potash	16.30	51.60	34.61
Soda	0.00	8.80	2.20
Lime	9.80	24.60	16.85
Magnesia	1.50	13.40	5.47
Oxide of iron	—	3.20	1.40
Phosphoric acid	9.20	22.60	16.80
Sulphuric acid	0.00	12.20	3.59
Silica	10.30	26.10	16.36
Chlorine	1.00	7.00	3.19

Ewell has calculated the following data in regard to the fertilizing constituents contained in the hop plant from analyses contained in the second part of Wolff's "Aschen Analysen"

ANALYSES SHOWING THE FERTILIZING CONSTITUENTS CONTAINED IN THE HOP PLANT (female) AMD ITS VARIOUS PARTS, STATED IN PARTS PER 100 OF THE AIR-DRIED MATERIAL.

	Ash.	Nitrogen.	Potash. (K_2O)	Phosphoric acid. (P_2O_5)
In hops	6.33	3.22	2.45	1.18
Leaves	10.50	3.46	2.01	0.36
Stems	3.12	1.57	1.08	0.23
Whole plant	6.49	2.50	1.59	0.38
"Spent" hops	3.54	3.33	0.53	1.23

FIG. 24. HOP FIELDS NEAR COOPERSTOWN, NEW YORK.

CHAPTER V.
THE CLIMATE AND SOIL FOR HOPS

FIG. 25. TANK FOR DIPPING HOP POLES TO PREVENT ROT.
Steppington hop farm, near Canterbury, Kent, England.

THE hop abhors continuous heavy fog or too much humidity in either air or soil, yet so rapid a grower must not suffer for want of water. Light fogs two or three times a week seem to favor hops, and to them Flint attributes the fine color so characteristic of Pacific coast hops. Winters that kill the root stocks are unfavorable. A climate that allows the root to rest from its labor, but enables it to make an early start in spring without danger from late frosts, an atmosphere free from excessive clouds and humidity, with abundant sunshine, not too dry as harvest approaches, with an absence of early frosts—there the hop thrives and there blights, mold and lice are reduced to a minimum.

Hence the superiority of certain limited regions in California. Oregon and Washington are apt to have too much moist, hot weather toward harvest. New York's climate is quite favorable on the average of years, but winterkilling is common. A climate in which corn (maize) does its best, is, in the United States, about right for hops, but, as Clark truly says, a great many soils and climates that are good for corn are bad for hops. English yards suffer most seriously from too much atmospheric moisture. The same is often true in Europe. Yet the tables of humidity, temperature and precipitation afford no guide to climatic adaptability to the hop. Yield, quality and price fluctuate quite regardless of meteorological statistics. We have spent much study over this point, comparing domestic and foreign weather figures with crop data, but without being able to draw therefrom conclusions of any practical value.

THE BEST SOIL FOR HOPS

This important subject has been considered for the present work by Prof. E. W. Hilgard, director of the California experiment station, whose knowledge of soils is not excelled, and who writes as follows:

California—"As to Sacramento county: Hops are grown almost wholly on the higher alluvial lands of the Sacramento river, which are gray, pulverulent, silty or sandy lands, with scarcely any noticeable change from soil to subsoil for several feet. Most of these lands lie near the river, where the land is higher than farther out; but some of the 'bench lands' beyond the overflowed region also yield excellent hops of the yellow silky character, while low-lying lands, not so well drained, yield a green product, which is less valued in commerce.

"As to Sonoma county and a portion of Mendocino to northward, the hop-growing lands are in the main the higher alluvial lands of the Russian river, greatly resembling in their nature those of the Sacramento just referred to; they are grayish, silty soils, uniform to several feet depth, well drained and of high fertility. The town of Hopland in southern Mendocino on the Russian river is one of the prominent growing centers, yielding a very high quality.

"In Alameda county only a small area is devoted to hop culture. It is located near the towns of Pleasanton and Sunol on the alluvial lands of Alameda creek, which are likewise of a fine sandy or silty character and well drained, as there is but little water in the stream beds in summer, and their banks are high.

"The oldest hop-growing region in the Pacific northwest is the valley of the Puyallup river in Pierce county, Washington. Here also the soils are alluvial ones, of a sandy or silty nature, of gray tint, very easily tilled and of considerable depth above bottom water, say from seven to ten feet. The Puyallup bottom was originally quite heavily timbered.

"The lands where the hop is grown in King county, Wash., lie on the lower Cedar and Dwamish rivers, and to northward on the borders of Lake Washington to the Snoqualmie river. Like all the lands of the Puget Sound region, these lands are of a light and sometimes sandy nature; the sand consists of the pulverized rock of the Cascade range adjacent."

In *Oregon,* the hop lands of the Willamette valley generally are light yellowish loams of great depth, and even the alluvium of the streams, like the Santiam, bears much the same character, though commonly lighter in texture than the lands of the main valley. It is conceded in Oregon that soil of a sandy nature produces the best quality, while the heaviest yield is to be obtained from the heavier bottom lands composed of decayed vegetation and deposits of sediment, brought down from the uplands and spread over this soil by the overflowing of the streams. The most perfect soil is a sandy loam which is easy to cultivate and is rich enough to produce a good crop of choice hops without the aid of fertilizers.

In *New York State,* and indeed everywhere, a deep sandy loam is preferred, the deeper the better for a crop with such a deep-growing root system. A clayey loam is also excellent if it contains enough sandy loam to prevent baking and packing during drouth. A strong loam in which corn thrives is generally good for hops, provided it is well drained. Its shallow root system enables corn to do well over a subsoil that would be too wet for hops, which also dislike too much gravel in the soil or a hardpan subsoil.

In *Great Britain,* the variation and yield in quality of hops in different soils, even between adjoining fields, is often most marked. This is equally true in New York state, Otsego and Schoharie counties usually producing the best hops. In New York, as in England, the lands now under hops have proven to be the best after centuries of hop-growing. The limits of the English hop lands are sharply defined geologically.

In the finest East Kent region, says Whitehead, the soil is clay, loamy clay, and sandy loam upon the Thanet, Woolwich and Oldhaven beds, which crop up here and overlie the chalk on the backbone of Kent. As the chalk appears again with a thin and gradually decreasing surface of loam, the hop land becomes less valuable, and at a short distance from this point hops are not cultivated at all until the bastard East Kent district begins, where the hops produced are of inferior quality as compared with East Kent hops proper, being grown upon useful, somewhat heavy soils, lying for the most part upon the belt of gault alternating with the Folkestone beds intervening between the chalk and the weald clay. Below Canterbury there is a district between Challock and Barham where hops of first-class quality are grown, upon loams of a lighter character resting on the chalk. The crops here are not so heavy as those yielded on the deep loam and brick earth in the Faversham district of East Kent, and the plants will not take such long poles, but the quality is most excellent. The "weald of Kent" is so named because of its soils resting largely on the geological formation called weald clay; they are clayey loams, sandy clays, more or less tenacious and stiff (these latter require expensive drainage), with occasional patches of loam and alluvium.

So, too, in *Germany*, the hop is more grown on clayey soils, well drained, than the average American planter would think possible. In Saatz and other famous Bohemian districts the soil is a reddish clay containing considerable iron, elevated about 800 ft above sea level and protected from cold north winds.

LOCATION OF A HOP YARD

Let it be naturally protected against prevailing wind storms, especially from the north and west. A heavy wind will badly whip the vines. (See "lewing," in Chapter X.) Very often this point is quite neglected in setting a hop yard, when it might just as well have been attended to.

Of course the site must be sunny and warm, and chosen with reference to the least possible danger from early and late frosts. The rows should run in a southerly direction, that the sun may freely penetrate the foliage to the utmost extent.

FIG. 26. HOP PICKERS IN WASHINGTON.

The main root is a deep feeder, its lateral and surface roots covered with fine rootlets that utilize the food in the upper layers of soil. Hence the need of a well drained soil—the hop abhors wet feet—and a soil of open texture, that air and water may freely penetrate, to aid in rendering available to the plant the elements stored up in the earth. Yet so gross a grower must have a sufficiency of moisture and drouthy lands may well be provided with irrigation.

PREPARATION OF THE SOIL

for a new hop yard is a more serious matter where the soil is not of just the right character. In Kent, expensive underdraining is often necessary to insure the needed openness of subsoils. Comparatively light yields in New York and in Germany are partly due to a moist or impacted subsoil. In such lands, thorough subsoiling to a depth of 18 inches, or even more, should precede planting. It is not much practiced, but is to be highly recommended. If subsoiling is needed for the sugar beet, which is dug in one season, how much more is it needed for the hop, whose roots go much deeper, but are not disturbed for from six to twenty years, or longer?

FIG. 27. PICKING HOPS IN KENT, ENGLAND.

59

It has also been suggested that subsoiling between the rows in early spring would be an admirable way of rejuvenating an old "root-bound" yard, at least on heavy soils. But Clark, speaking for New York conditions, says: "I disagree very strongly with subsoiling between the rows, or even deep plowing of an established yard, as that space is filled with large bed roots, and deep culture cuts them off, which is very injurious. I have seen them 16 feet long in my own yard. Ottenheimer says that for the Pacific coast, plowing deep when setting out the yards is right, but afterward it is injurious to subsoil each spring."

The tendency is also to slight the surface plowing for a new hop yard, just as thorough working of the soil preparatory to seeding down to grass for several years is too commonly neglected. While experts differ as to the propriety of putting the plow into a yard once it is well established, every intelligent grower realizes that before the roots are set affords the best chance to thoroughly work the soil. The English realize this and practice accordingly in preparing for hops, just as they do in preparing for the permanent meadows for which old England is famous. The Germans are not so particular.

CHAPTER VI.
FEEDING THE HOP PLANT

THE hop is a rank feeder. The most of its growth is made in less than 90 days. This growth is marvelous for its luxuriance. Such luxury of foliage is necessary if the hops are to have a copious supply of properly elaborated elements in the plant to draw upon during their maturity. The plant must be fed for growth as well as for fruit, the one being dependent upon the other, but avoid such treatment as will force it to "run to vine" too much. These points have only to be recognized to realize the necessity for proper soil, appropriate fertilization and correct methods of culture.

FIG. 28. PICKING HOP.

Of course a virgin soil filled with fertility, or renewed by an annual overflow or by irrigating with water naturally rich in the elements of plant food, requires little or no manuring. Such is the present condition of many of the newer yards on the Pacific coast, but it is only a question of time when even they will require manuring. How best to feed the hop on the more or less exhausted lands of the eastern states and of the old world is a problem upon which we have comparatively little exact data. The experiment stations in Bohemia are attacking this problem, likewise the Wye Agricultural College in Kent and a little has been done in Germany, but American experiment stations seem to have largely ignored the problem of fertilizing the hop. Let us, then, first consider the elements of plant food contained in the vines and hops of an average crop, basing our table on the analyses on this page and on the average relative weight per acre of vines and hops obtained from a dozen experienced growers in New York state:

A GOOD CROP OF HOPS WILL TAKE FROM AN ACRE OF LAND

	Cured hops	Vines and leaves (air-dry)	Total.
	lbs.	*lbs.*	*lbs.*
Weight of crop	1,000	1,000	2,000
Nitrogen	33	25	58
Potash	25	19	44
Phosphoric acid	12	7	19
Lime, magnesia & other ash elements	30	42	72
Tot. removed by crop	100	93	193

These are astonishing figures. Their significance can be best judged by comparison with the plant food removed from an acre by other crops under equally good culture, it being assumed that the hop vines, like potato vines and cornstalks, are returned to the soil:

PLANT FOOD REMOVED FROM AN ACRE BY SEVERAL CROPS

Crop.	Hops.	Hay.	Corn.	Potatoes.
Yield per acre	1,000 lbs.	1½ tons	40 bu	250 bu
Nitrogen, lbs	33	42	41	30
Potash, lbs	25	45	10	45
Phosphoric acid	12	8	16	11

How few hop planters in New York state realize that for a good crop of hops they must manure as heavily as for 40 bushels of corn per acre, simply to supply what is taken from the soil by the dry hops. If we consider both vines and hops, we get this table, showing:

COMPOSITION AND QUANTITY OF MANURIAL SUBSTANCES
REQUIRED TO SUPPLY WHAT AN ACRE OF HOPS
TAKES FROM THE SOIL

Pounds.	Substance.	Will furnish		
		Nitrogen.	Potash.	Phos. acid.
		lbs.	*lbs.*	*lbs.*
2,000	Hop crop	58	44	19
2,000	Wheat bran	52	32	60
1,000	Cottonseed meal	70	20	30
1,000	Linseed meal	55	14	17
5 tons	Barnyard manure	50	40	30
100	Bone meal	4	—	23
100	*a* Wood ash	0	50	10

a This weight of wood ash (containing only 12% water) will supply the full amount of potash taken off by the hop crop (vines and hops) but no nitrogen; the other weights given will furnish the full amount of nitrogen, but more or less of potash and phosphoric acid than the crops take off, except in the case of bone meal.

Stable manure is the form of plant food preferred by both European and American hop growers. In compact soils horse manure is best, because of its mechanical effect in lightening the soil as well as furnishing food to the plant. Sheep manure is excellent for sandy soils. Ordinary mixed stable manure is plowed under lightly in starting a new yard, when the soil is at all poor. The amount should be all one can possibly afford, and then a little more; no danger of getting on too much before planting a new yard. In New York from 10 to 20 tons per acre of stable manure are applied in starting a new yard, in England 15 to 25 tons, and in Germany eight to 18 tons.

After the yard is established, fall application of stable manure is best on most soils. The common practice is to put a shovelful or two of manure on top of each hill in late autumn, to be scattered about the hill and worked into the soil at first grubbing in spring. Green (fresh) manure should not be used, as it holds the frost too long in spring, will not work readily into the soil, and interferes with cultivation. In cold regions this protects against winterkilling, and in case of drouth protects the roots by retaining moisture. The manure washes down about the roots and aids a prompt and early start, but if cold weather follows, this may result in stopping the flow of sap and arresting the growth of the plant. On very sandy soils, such dressings may be quite exhausted before the plant blossoms out, and the vine has no reserve of fertility with which to develop its hops.

In addition to this autumn manuring on the hills, a dressing of manure broadcast is highly recommended, to be worked into the soil at the first cultivating. If the soil is very light and leachy, broadcast the manure in early spring, but if fairly strong loam, midwinter spreading is best. In Germany a liberal mulch of strawy manure is often applied after cultivating is finished, especially on drouthy lands. Its preservation of soil moisture is quite as useful as the food it furnishes the plant. Such a dressing must not be so rich as to cause the plant to run to vine to the detriment of its production of hops.

Comparison of the analyses printed on Page 62 with the analyses of Kent Goldings and Sussex Grape hops grown in England, shows wide variation in the total per cent. of ash of vine, leaves and cones between different varieties and even the same variety grown on different soils. We find no analyses to indicate the variation caused by different forms of plant food, but it is quite probable that the influence of the form of food upon the hop plant is more noticeable in its brewing qualities or its organic composition than in the proportion of ash or nitrogenous matter. This is an extremely interesting point upon which scientific experimentation will doubtless throw much light. On general principles, however, it would seem expedient to employ the least objectionable forms of plant food, when agricultural chemicals or commercial fertilizers are applied.

Potash is needed to excess, owing to the great demands upon this element by the plant, and probably the carbonate of potash, as in cottonhull ashes or unleached wood ashes, is for many reasons preferable. Of the potash salts, the high-grade sulphate, which is much freer from chlorine than the muriate, is perhaps best. Yet, there is a large amount of chlorine in the hop, and should it be scientifically demonstrated that the presence of a liberal amount of this element was essential to certain desirable qualities, then the muriate of potash would be used.

There seems to be little reason for believing that one form of phosphoric acid is much better than another for the hop crop, provided only that it is in a form that will be available for the plant. Bone and ashes furnish both phosphoric acid and potash, but in a slow form, and as the hop is a rapid grower, and requires an abundance of available food early in the season, it is probable that the application of potash salts and dissolved boneblack or other quick-acting phosphate would be beneficial. This quick fertilizer should be applied very early in the spring, while bone and wood ashes should be put on in the fall.

COMPOSITION OF HOP MANURING SUBSTANCES

The figures show the per cent. or pounds of each element in 100 pounds of the substance named in first column.

Substance.	Nitrogen	Potash	Phos. acid	Lime	Magnesia
Wheat bran	2.6	1.6	3.0	0.2	0.9
Cottonseed meal	7.0	2.0	3.0	0.3	1.0
Linseed meal	5.5	1.4	1.7	0.4	0.8
Rape meal	5.1	1.3	2.0	0.7	0.7
Barnyard manure	0.5	0.4	0.3	0.5	0.1
Bone meal	4.0	0.0	23.0	31.0	1.0
c Boneblack dis'lved	0.0	0.0	17.0	25.0	0.7
c Phosphate rock, dis.	0.0	0.0	15.0	23.0	0.0
Tankage	6.7	0.0	12.0	14.0	0.0
Dried blood	10.0	0.0	2.0	0.8	0.2
Wood ash unleached	0.0	5.0	2.0	34.0	3.4
Cottonhull ash	0.0	22.0	8.0	10.0	11.0
Kainit	0.0	13.5	0.0	1.2	10.0
a Muriate of potash,	0.0	51.0	0.0	0.0	e. 0.0
c b Sulphate of pot'sh	0.0	33.0	0.0	0.2	e. 0.0
d Nitrate of soda	15.7	0.0	0.0	—	—
c Sulphate of am'nia	20.5	0.0	0.0	—	—

a Contains 48% chlorine. b No chlorine. c Rich in sulphuric acid. d Contains much soda. e Traces.

The hop is a great consumer of lime, yet the application of lime to hop yards is comparatively rare. We see no reason why it is not advisable, unless the soil is known to contain an excess of lime. This element is equally important in tobacco culture, where the use of lime is considered indispensable. Probably the best form is oyster-shell lime, provided it can be obtained at a nominal price. Otherwise, good air-slaked lime can be used, or the fine ground gypsum (land plaster); from 100 to 300 pounds of lime per acre, applied in the fall, is sufficient, usually. Most soils probably contain sufficient soda, but if not, it is a prominent composition of many potash salts.

It may be desirable to add magnesia to some soils, in which case the double sulphate of potash and magnesia should be used instead of kainit, muriate or sulphate of potash alone.

Perhaps the most striking need of the plant is for nitrogen. We have no data to show to what extent, if any, the hop plant is able to take its nitrogen from the atmosphere, as do certain leguminous crops. But we do know that it is a gross consumer of nitrogen and that this element must be in a promptly available form to promote the vine's luxuriant growth. Hence, the importance of applying nitrate of soda, or sulphate of ammonia to give the crop a quick start in spring, and some less soluble form of nitrogen to back up the crop as the season advances, such as dried blood, tankage, or bone meal.

In this country a few manufacturers of commercial fertilizers have attempted special mixtures of agricultural chemicals for the hop crop, with more or less success. In England special hop fertilizers are far more common. We cannot recommend any one formula as the best for this crop in different soils, but the following table contains the composition of various fertilizing materials, and from the known composition of the hop plant, several mixtures are tentatively suggested:

Formulas for Manuring Hops

It is fair to assume that, provided the vines are carefully returned to the soil, 1,000 lbs. per acre of cured hops will remove plant food varying within the range below stated. And to supply either of these would require the mixtures which follow:

In 1,000 lbs. of cured hops. Formula.		Least quantity.				Largest quantity.		
		Nitrogen	Potash	Phos. acid.	From.	Nitrogen	Potash	Phos. acid.
	In 1,000 lbs. of cured hops.	25	20	9		38	28	14
1	Stable manure.	2½ tons	25	20	15	8 tons	40	32	24
2	Cottonseed meal	400 lbs.	28	4	12	600 lbs.	42	6	18
	Kainit	200 lbs.	0	27	0	200 lbs.	0	27	0
	Totals	**600 lbs.**	**28**	**31**	**12**	**800 lbs.**	**42**	**33**	**18**
3	Linseed meal	500 lbs.	27	7	8	700 lbs.	38	10	12
	Cottonhull ash	60 lbs.	0	13	4	100 lbs.	0	22	8
	Totals	**560 lbs.**	**27**	**20**	**12**	**800 lbs.**	**38**	**32**	**20**
4	Bone meal	100 lbs.	4	0	23	100 lbs.	4	0	23
	Dried blood	100 lbs.	10	0	2	200 lbs.	20	0	4
	Sulphate potash	40 lbs.	0	20	0	100 lbs.	0	33	0
	Nitrate soda	50 lbs.	8	0	0	75 lbs.	12	0	0
	Totals	**290 lbs.**	**22**	**20**	**25**	**475 lbs.**	**36**	**33**	**27**
5	Sul. ammonia	100 lbs.	20	0	0	150 lbs.	30	0	0
	Wheat bran	200 lbs.	5	3	6	300 lbs.	8	5	9
	Wood ash	200 lbs.	0	10	4	200 lbs.	0	10	4
	Muriate potash	20 lbs.	0	10	0	30 lbs.	0	15	0
	Totals	**520 lbs.**	**25**	**23**	**10**	**680 lbs.**	**38**	**30**	**13**
6	Phosphate rock	100 lbs.	0	0	15	100 lbs.	0	0	15
	Sulphate of potash	60 lbs.	0	20	0	80 lbs.	0	26	0
	Linseed meal	100 lbs.	5	1	1	200 lbs.	10	2	1
	Nitrate soda	50 lbs.	8	0	0	100 lbs.	16	0	0
	Sul. ammonia	25 lbs.	12	0	0	60 lbs.	12	0	0
	Totals	**335 lbs.**	**25**	**21**	**16**	**640 lbs.**	**38**	**28**	**16**

Many other combinations of the ingredients mentioned on Page 67 may be made. But in any formula, the object should be to supply the nitrogen, potash and phosphoric acid in such forms that part of each element shall be available for the plant in early spring, and then from week to week, as growth advances, but not force a growth when the plant is maturing its cones.

The large proportion of nitrogen contained in hop vines is wholly lost when they are burned, though the mineral elements are retained in the ash. Since this plant draws so heavily upon soil (or air) for this most expensive element, certainly it should be retained so far as practicable by plowing under the vines, provided they are not infested with germs of disease so as to require burning. Spent hops are specially rich in nitrogen, and when they can be had for the hauling, should be spread on the ground and cultivated under.

FIG. 29. IRRIGATING HOPS, MAKING A LITTLE WATER MOISTEN MANY ROWS OF PLANTS.

FIG. 30. A YARD ON THE SHORT POLE SYSTEM.

At Watsonville, Santa Cruz Co., Cal. Poles are. 2x3 inches x 9 feet long, of split redwood, set 2 ft. in ground, 8 ft. apart square. No wire is used, only No. 18 cotton twine, which is fastened to pole 6 ft. from ground. The string is run in squares, and two vines are left to the hill. Vines are trained on poles up to the strings. Mr. Morse allows one male vine to every 35 female vines; males are not pruned and are given 15 ft. poles to climb. They consequently grow very bushy, and, as they climb to the tops of the high poles set for them, a good distribution of pollen is secured. This short-pole system is not to be confused with either the trellis system of overhead wires, or the long-pole method used in Washington, New York and England.

A great number of other substances are much used in England and on the continent, such as shoddy, waste, woolen rags, fur waste, fish manure, and basic slag from phosphoric acid. Irrigation may here be practiced, for it is essentially a feeding process. No matter how much plant food is in the soil, unless there be sufficient moisture, the crop cannot utilize it. Moreover, the hop must have an abundant supply of water, because nearly nine-tenths of the vine's weight consists of water. Frequent stirring of the top soil, or a mulch of strawy manure, leaves, weeds, cornstalks, or any such material will carry a crop through a drouth that would otherwise be fatal. Where irrigation is practiced in California, the water is run through one furrow in the middle of the rows, or one on each side. Sometimes two or three such irrigations are enough, again more may be necessary, while in a Colorado hop yard, the water is turned on six to nine times. If the water supply is scant, a very little can be made to irrigate a large number of plants by the device illustrated in Fig. 29.

CHAPTER VII.
Laying Out a Yard—Training the Vines

HOP plants are usually planted 7x7 feet or 8x8 feet in America and 6x6 feet in England and Europe, but the number of hills may vary from 800 to 1,200 per acre. In New York state 6½ feet each way is preferred by some experts.

On the Pacific coast in very few yards are hop roots planted less than seven feet apart, and in a great many yards the rows are eight feet apart. It has been demonstrated there that just as heavy a yield can be obtained from a yard planted with the roots seven feet apart as from one 6 or 6½ feet apart, notwithstanding that in the former there are only 889 hills to the acre, while in the six-foot yard there are 1,280 hills. As the most expensive part of raising hops is the work done by hand on each root and vine, such as grubbing, tying and training, it can be readily seen that the expense to cultivate an acre of hops is considerably larger in a six-foot yard than in a seven-foot one. Where the trellis system is used, it requires a great deal more twine in the six-foot yard. Another objection is that a team of horses cannot pass through a six-foot yard without injuring the roots or vines. In Oregon, as well as in Sonoma county, California, nearly all the hop yards are set out with the hills eight feet apart.

The method of laying out the yard is therefore much the same everywhere, though the methods of training the vines are almost "too numerous to mention." It is important, in any system of training, that the rows be perfectly straight to facilitate clean culture. Now, let us assume that the field is ready for staking out.

Set plain, distinct posts at the four corners of the plat; then take a long wire with a stake at each end, and at a distance of every seven feet tie a piece of flannel or cloth, to be easily seen. From one corner stake, sight in a direct line to the other corner stake, pull the wire tight and firmly set in the ground. Put in a peg about a foot long at each marker on the line and then again continue the line in the same way, pegging until that side is pegged. Next, from that corner stake and at right angles, take the side to the corner stake at the other end of that side, as above described, pegging as you go on. Then from each of these outside rows of pegs, start to the other side, having set up a stake to sight to, seven feet distant each time. When both sides are thus completed, the field will be pegged out as illustrated in Fig. 43.

FIG. 31. TRAINING HOPS IN KENT.

The land may be marked off similarly by a variety of means. Mr. Clark writes: "Make a marker in the form of a bob sled, with short runners of one and one-half inch ash with a light shoe. Bore a hole through the runners about a foot from the back end and about two inches from the lower end, so as to be able to put clevices in to help make better marks. The top of the marker should be made of stout 1¼ inch ash boards for the driver to stand upon. Place an iron handle on the center of the back so as to help in lifting the marker around at the ends of the field to the center of the front side. Fix an upright standard about four feet high for the driver to take hold of; it will also serve as a guide. After fixing on a pole for a pair of horses the marker is ready. It is a good plan to have a couple of boys standing at about equal distances across the hop yard with flag stakes so that the driver, when standing on the marker, can look between the horses' heads and see the stakes. By so doing, he can make two very straight marks and also get over the ground very rapidly. The land should be marked both ways, but never with a plow, or one row will be narrow and one wide."

FIG. 32. HORIZONTAL HOP YARD, NEW YORK.

If poles are used, one or two poles are inserted at each hill; the single pole is now most common in New York, but two poles per hill are much used abroad, leaning outward from each other so the hops will not mass together at the top. Cedar poles are most durable; in Washington they are sawed out or split about 3x3 inches, 16 feet or more in length, for the long pole system, and 10 feet long for the short pole. An eight-penny nail is driven in the top, projecting out about an inch, in the short or stake system; on the long pole, about a foot from its top, put through a peg a foot long and three-fourths of an inch thick for the vine to cling to. This and the sharpening of the poles is done in the woods or at the mill. In California and Oregon the poles are split in the same manner as fence rails. Redwood poles are quite extensively used in California, and they last an indefinite length of time. In some yards the same poles have been used for the last 25 years. In Oregon those growers who adopt the pole system use young firs, which grow abundantly in that state. They aim to get a pole three inches thick and about sixteen feet long. In New York and abroad, round poles are used, from saplings, and are not as high as those on the coast.

Stand the poles upright in a tank containing two feet of creosote or coal tar, and let them simmer over a slow fire for a night; this will prevent the butts from rotting and is a big saving. Cedar, ash, redwood, chestnut, maple, oak, alder, and birch are esteemed in the order named for hop poles.

On the Pacific coast, when a crop is picked the first year, poles are set before the roots are planted, which prevents injury or disturbing the roots afterward. With a long dibble having a steel sharpened point, a hole is made, about eighteen inches deep, into which the pole is stuck and left vertical. A man will set about 600 poles per day. A short stake is set the tenth hill in every tenth row to indicate when a male root is to be planted.

In New York, England and Europe, poles are not set until the second year, care being taken to set the poles in the north side of the hill every time, as the men cultivating before the hops are up will know better where the hills are and will not be so apt to damage them. In later years also the men when setting the poles will know better where to find the old holes. For a short pole yard, the stakes could be cut (for economy's sake, split) ten or more feet long, that they may be long enough to use after once rotting off. The outside rows should have larger stakes and be set very deep and solid.

Twine is run across the top of the poles both ways, being attached to the nail, or some merely wind it around the poles at a height of 7 to 7½ feet from the ground (Fig. 32). After the first year, not more than four vines should be trained to each hill by this system, and where the soil is extra heavy, two will be found preferable. At the first and second trainings, all surplus vines should be either pulled out or cut off beneath the surface. The vines generally require training twice before reaching the twine, and the vines should be trained at least twice on the twine. In training on the twine the first time, it is best to take the vines from the stake above the twine, and after passing them across over one twine, bring them down under the second twine and train out on the second twine. This causes the vines to arch over the twine and prevents them from pulling down on the twine next the stake, thereby preventing the twine from either breaking or stretching, which would cause the hill to slide to the ground when heavy with the weight of full-grown hops.

FIG. 33. OTSEGO (N. Y.) GRUB HOE.

There are several modifications of the short pole and twine method. A popular one consists of driving a nail (slanting downward) into the pole only about four feet from the ground, tying the string to the top of the next pole, and so on. Drive the nail first into the first pole in the first row, then go to the second hill in the opposite row, then back to the third hill in the first row, and so on across the yard, doing two rows at once. Begin by tying the twine to the first nail, run the top of the twine up the next pole with a "twiner," as far as convenient, carrying it around the pole and trying to catch the twine over a knot to hold it; draw up the twine close, then drop from the top of the pole down to the nail in the next pole. Step up to it and give the twine a half hitch or loop around the nail, then run the twine up to the top of the next pole, down to the next, and so on across the yard, until all are finished in the same way. Then turn and go across in the same manner, getting the effect shown in Fig. 3. By this system, Clark claims that more hops can be grown, they will mature earlier, be richer and brighter, will arm out lower down, and the arms will be longer and not apt to snarl up. They will fill up in the middle with soft, white, undeveloped hops and will make better picking, and are not as leafy.

FIG. 34. TWINE POLE.

Still other modifications of the twine system are used in England and on the continent, which are sufficiently explained in the accompanying illustrations. By whatever method twine is used in these systems, a device for tying the string about the poles is useful. It consists of a strong but light pole, eight to 12 feet long, with screw eyelets like a fly rod, and a bag or basket at the bottom that will hold a ball of twine snugly (see Fig. 34) A good 12-ply cotton string is used. More permanent methods of training by means of wire trellises are constantly coming into wider use. The first cost of these methods is more than for the pole and twine system, but where hops are grown on a large scale, some form of wire is probably the more economical. It is claimed also that the hop vines can be kept open to the sun more thoroughly by trellises than by the string system. There is also considerable saving in labor, after the method is once established. Spraying can also be done more thoroughly when the vines are spread out on proper trellises than when they grow more closely together, or simply on poles. Again, the hops are not wind-whipped as readily; it is claimed that they mature earlier, can be picked cleaner, and come down in better condition.

Whitehead says: "One arrangement of wires and string is much adopted in East Kent. It consists of stout posts set at the end of every row of hop stocks, and fastened with stays to keep them in place. At certain intervals in each row a post of similar size is fixed. From post to post in the rows wires are stretched at a height of half a foot from the ground and at a height of six feet from the ground, and again from the tops of each post: so that there are three lengths of wire in all. Upon these wires, hooks are fastened or 'clipped' at regular intervals, so that cocoanut fiber string can be threaded onto them horizontally from the lower to the next wire, and in a vertical direction from this wire to the top lateral wire of the next row. The string as threaded on the hooks is continuous, no knots are necessary, and it is put on the hooks of the top wires with a 'stringer.' The first cost of this is about $200 per acre."

Another method is that shown in Fig. 35, and practiced extensively in England and Germany. By this method, wires are fastened only to the tops of the posts, and twine is run down to pegs in the ground, these being more simple and less expensive than the system just described. The stay pole, or what the English call "the dead man," must be very firmly set and the end pole braced to it by wire. In New York this method is further simplified by setting poles 18 to 20 inches deep every sixth hill, running a single wire along them from nine to 15 feet above ground, and two strings only running into a small, wooden or wire plug driven firmly near the hop plant. The latter idea has been still further improved upon by the Pleasanton Hop Company, Alameda county, California. As this concern is one of the largest hop growers in the world, and has made many improvements in the industry, we are fortunate in being able to devote Chapter IX to an exact statement of its *modus operandi*, carefully prepared for this work by Mr. Davis, superintendent of the Pleasanton Hop Company.

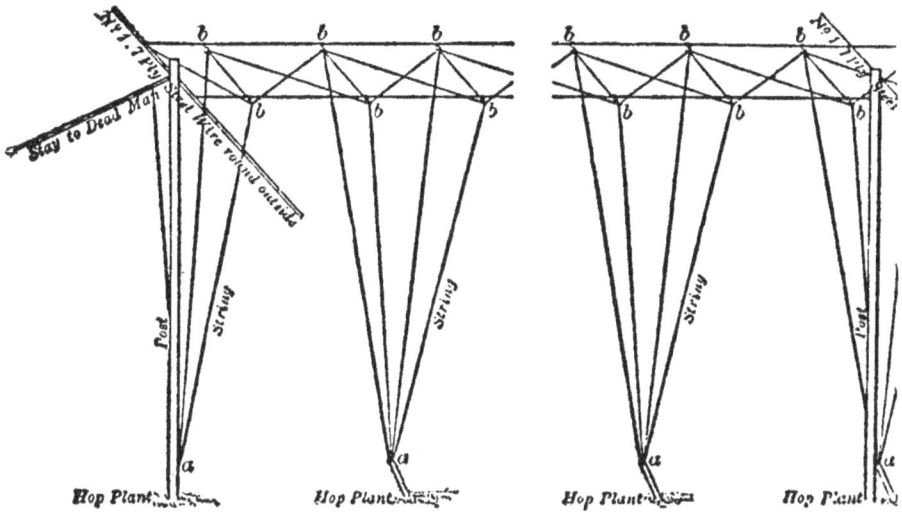

FIG. 35. KENTISH WIRE TRELLIS.

FIG. 36. KENTISH HOP YARDS.
Two poles to a hill and string.

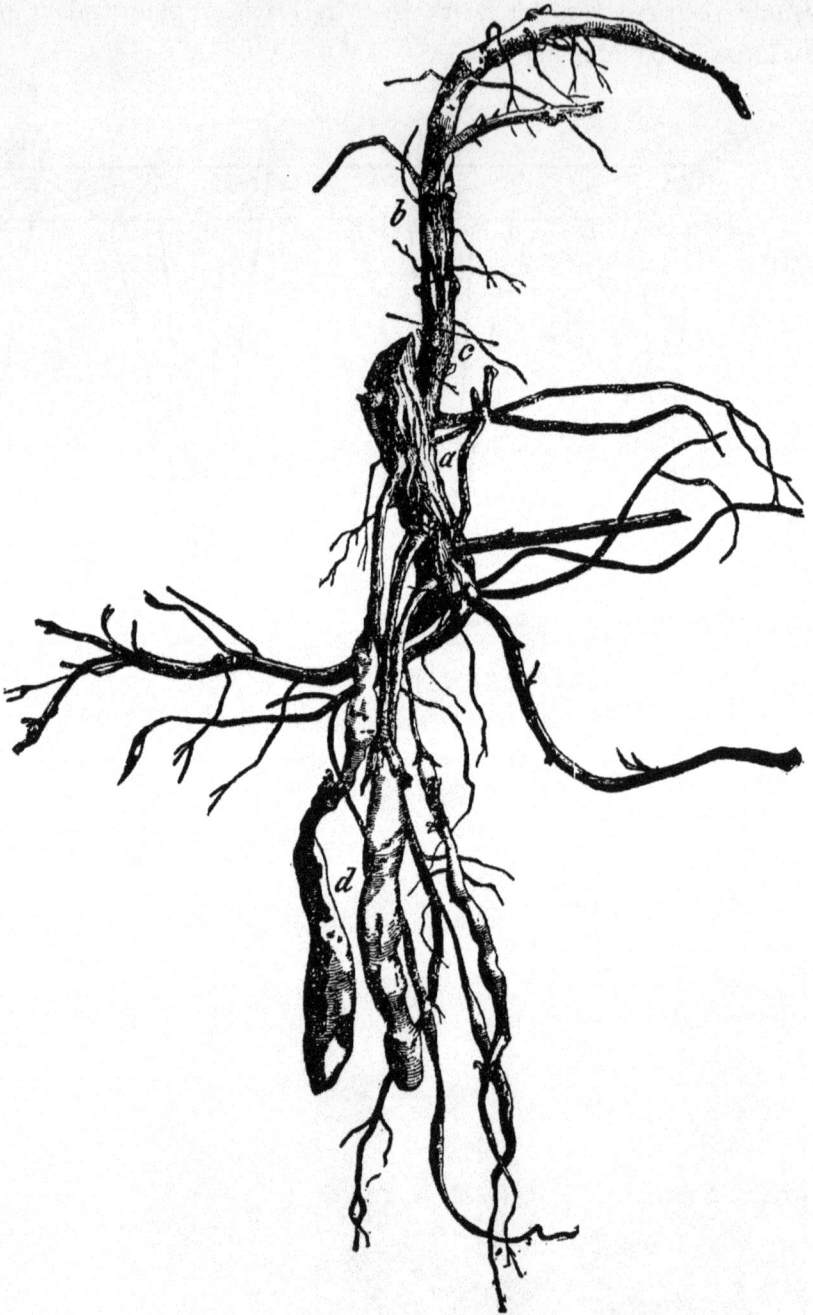

FIG. 37. THE HOP ROOT STOCK.

a, Main stock; *b*, base of last year's growth of vine; *c*, where the vine was cut off when the stock was trimmed up for the new growth; *d*, tuberous appendages.

CHAPTER VIII.
PLANTING AND CULTURE

HOP roots raised from seed are preferred by a very few in starting a new hop yard. The seed used should of course be of good quality. It should be perfectly ripe and taken from well-developed canes of vigorous specimens of the same species, all of which have arrived at maturity together. By so doing, there is a better chance of obtaining plants which will develop and mature at the same time. This point is important, as affecting cultivation and the expenses of picking the crop.

The seed should be sown in a hotbed frame, either broadcast or in rows; the soil should be light, well manured and thoroughly pulverized. If the seed be sown in rows, the spaces should be marked out with a line about two or three inches apart and about a quarter of an inch deep. If sown broadcast, the method in use in sowing all fine seeds should be adopted. The seeds should be covered by means of a rake, the back of which may be used to gently press down the soil, after which the teeth may be used to mellow the ground still further. The seed should not be covered by more than an eighth of an inch of earth. A little chopped straw should be spread over the surface in order to prevent the seed being exposed or the plants washed out when watered. Watering is indispensable to maintain the earth in a proper state of humidity for the germination of the seeds and for the development of young plants. When sown broadcast, the seed must be covered up with a rake and operated upon as if sown in rows. The seeds germinate in six weeks and the plants are ready for use the following month. They should then be put out, but will yield no crop until the following year.

Hop roots for planting are usually cut from old stocks or runners. From such cuttings fully 99 per cent. of the yards are planted. The hop roots should be cut into pieces from four to five inches long, with two sets of eyes on them (that is, two joints), one for the roots, the other for the vines. It is more accurate to say that the lower roots grow from the extreme lower ends, or from little pimples on the side. Great care should be taken to have the sets of strong constitution, in prime condition, and absolutely true to name. Before being planted, the sets should have their roots properly trimmed and dead growth removed. They are dug out a few days prior to planting, so as to get a trifle dry, to prevent them from bleeding to death when planted. Of course, they must not get dry enough to destroy the life in them, and it is also important that they be whole and good. In California's dry climate, the roots are set out as soon as ready, or "heeled in," to prevent drying. If fresh-cut sets are planted, sifting plaster (gypsum) over them is often done.

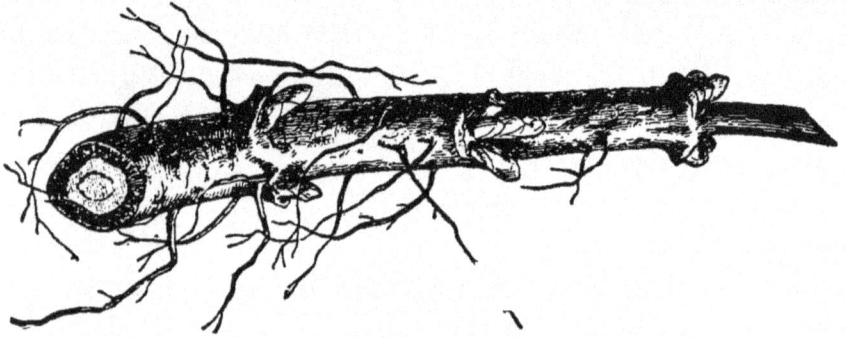

FIG. 38. HOPE VINE STOCK FOR TRANSPLANTING.

Opinions vary as to what constitutes the best root. On the coast, a root is preferred that is cut from near the outer end of the runners and the roots should all be of near the same length and free from split or bruise. The ends should be cut perfectly smooth, and each root should have not less than three or four sets of eyes and one set of eyes should be near the upper end of the root. In England and Canada, these cuttings are at once planted six inches apart in nursery rows two feet apart, the roots being removed and leaving only three or four eyes around the stalk. These stalks will make roots and a moderate growth of vine, and will be ready for transplanting the next fall or early spring. Each root will then be a crown; that is, one that has carried a vine, not a sucker. Sometimes, when sets are very dear, the pieces of root cut off in the spring are planted out at once without having been put in a nursery; this is frequent on the coast, but is not practiced at the east or abroad. Sets are more often cut direct from old roots.

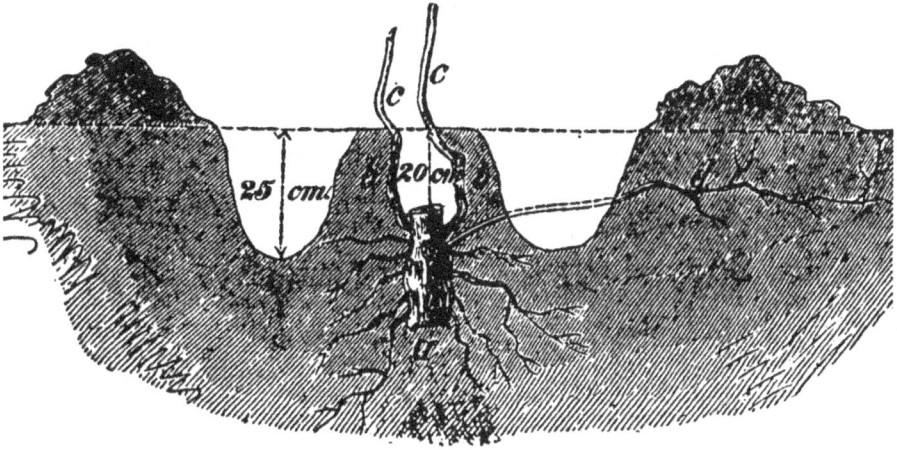

FIG. 39. THE HOP STOCK.

a, Rootstock; *b*, vine stock; *c*, dead part of last year's vine cut off; *d*, roots separated or cut off by grubbing.

There are several methods of planting the sets. One is to make a hole 14 inches deep, with a hop-bar, put in a handful of fertilizer, partly fill the hole with loose soil, and set the roots on end, with the top even with the surface of the ground, eyes sticking up and tops spread apart. This way of setting will produce a hill with the crowns close together, which is a great advantage over the old way of planting, where the crowns grow a foot or more apart. Besides, roots set in this way will not be affected by drouth, because most of the root is much deeper in the ground than planted ones. The old way is, if the pole was previously set, to dig out the soil about six inches deep on the south side, spread out the roots carefully, eyes up, cover to crown, about an inch deep, and level with the surface, and firm it with back of hoe. If covered too deep, the sets may smother.

In England, square holes are made with a spade, their exact center indicated by a stick, and the sets are pressed in firmly with hand and foot, an inch or two of the sets being left above ground, a practice that is not followed in this country, although it is to some extent in Germany. Some farmers plant the roots in the same way as potatoes by simply dropping them in the mark. But they are apt to dry up or remain dormant for a long time, in a dry season. They will thus get a late start and make a feeble growth; the runners, moreover, are apt to shoot out a good deal from the hill, thus getting the hill out of place.

In Washington, the yard is planted by shoving a spade deep into the ground on the same side of each hill peg, and setting two roots in the hole thus made, placing one in each side of the spade hole, and then pressing the soil firmly around them. Care is taken to get the roots set perpendicularly and the right end up, or on a slight angle pointing to center of hill, and the top end covered one or two inches below the surface and level of the ground.

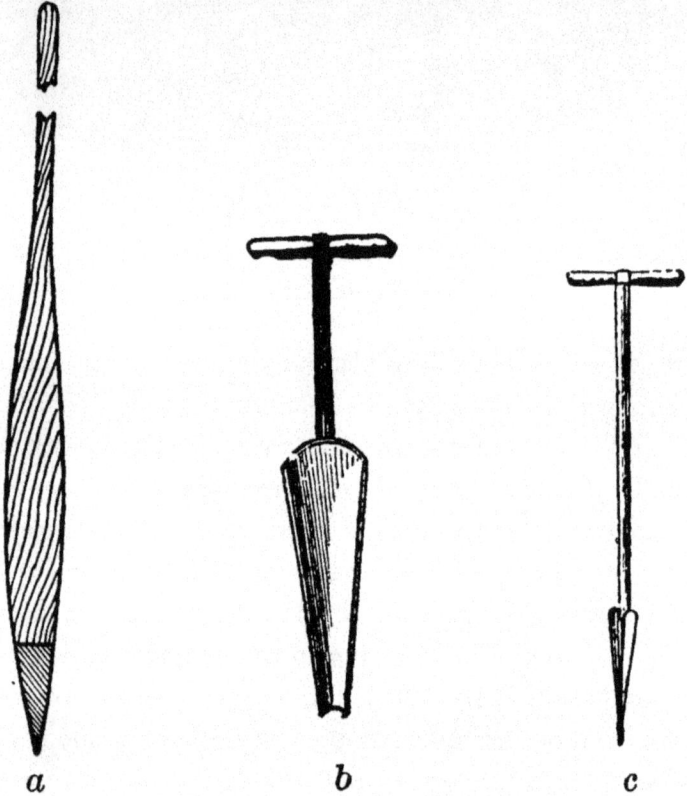

FIG. 40. TOOLS FOR MAKING HOLES FOR SETTING POLES
a, Wooden bar with iron point; *b,* hole augur; *c,* pointed dibble.

Planting should be done as early in the spring as possible, because the hills will thus get a much stronger growth and the crop of the following year will be from 25 to 50 per cent. larger than from late planted hills. Fall planting is the rule in England and on the continent. In California, planting is done in January and February, and further north during March and April. About three bushels of good roots may be allowed for planting one acre of ground.

There should always be three or four rows planted across the field (between the hop vines), with hills that are two feet apart. This will give a supply of extra roots that can be taken up the following spring to fill in the missing spots where some hills may have died out. A nursery should also be planted every year, so as to have sets to fill in any of the missing hills the following spring.

If the new plantation is not to yield a crop of hops the first year, poles are not set, but stakes 12 or 15 inches high are driven in the north side of each hill to mark its location. The space between the rows is then planted to some hoed crop. Corn is frequently grown, but is objectionable because of its heavy shade. Beans are better, because they do not shade the plants so much and do not rob the soil. Potatoes are often used also, or lettuce and other small crops are grown under intensive culture. The small marking stakes will do for the young vines to twine about. Clean culture is to be carefully pursued the first year. Weed out the hop rows and place a little fresh dirt around them, but do not work the hoe very deeply about the young plants. Even if the soil is rich, it is wisest not to grow any other crop the first year, and certainly not thereafter, for the plants will need all the fertility the soil contains. At least, it must be very liberally manured if a catch crop is raised the first year.

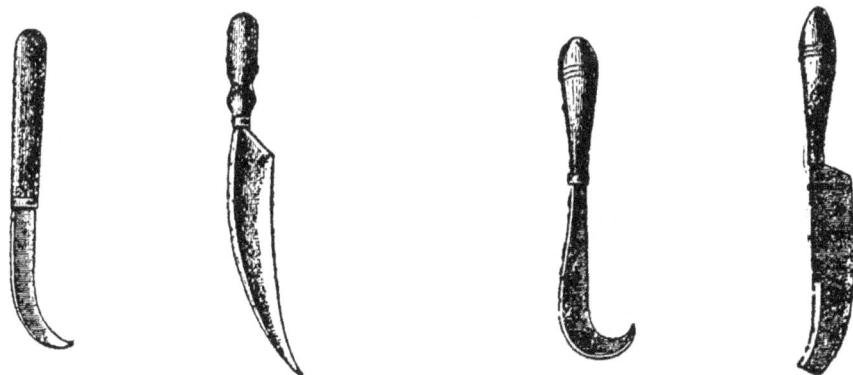

FIG. 41. FORMS OF HOP KNIVES.

If the new plantation is to be worked for a crop of hops the first year, its culture is practically the same as the treatment of an old or second-year hop yard after the grubbing out. See Chapter VI for particulars about manuring or fertilizing.

83

CULTIVATION DURING THE SECOND YEAR

In the spring of the second year, dress out the hills with a four-tined fork and work in the manure thoroughly, being sure to cover up the shoots, as both freezing and hot suns will do them considerable harm. Some plow the soil away from each side of the plants, but even when this is done with care, Clark and others protest against ever putting a plow into the hop plantation. They prefer to remove the fall dressing to one side of the hill, then with a grub hook (Fig. 44) loosen and remove the earth from around the hill to the depth of three or four inches, pulling up and trimming off the surface runners and cutting off (with knife like one of those shown in Fig. 41) the crown or top an inch or two, as shown in Fig. 42. The old manure is now worked into the soil about the hill and the plant covered with fine earth. The English use a special tool (Fig 45) for hauling the fine earth over the trimmed plants. Any dead roots must be replaced, also diseased or decaying ones. This is the proper method of "grubbing out" every spring.

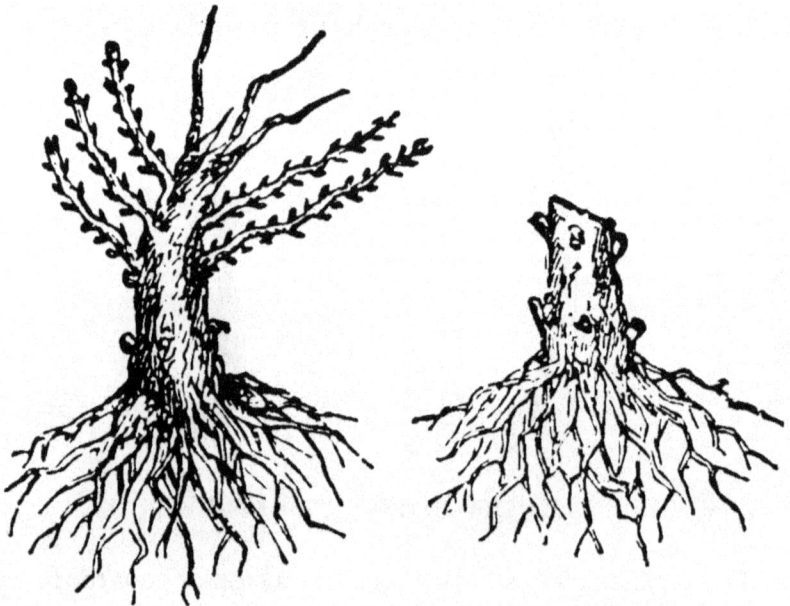

FIG. 42. PLANTS UNTRIMMED AND TRIMMED.

An important point is thus stated by Whitehead: "It is well not to 'dress' hop plants too early, as, if the shoots or bines are forward, they are exposed to the

84

action of spring frosts, which will either cut them up, or blacken and spoil them, or make them 'sticky,' unkindly, and more liable to blight and mildew. The French vine cultivators dread the influences of white frosts upon the young and tender shoots of the vines, which are most pernicious, especially if the sun shines on the vines while they are covered with dew. On the other hand, if the plants are dressed very late, and cold, dry weather comes in May, as is sometimes the case, the bines get behind and cannot make up for lost time. But most planters now hold that moderately late is better than too early dressing. Care must be taken in dressing not to cut the stocks too low, thus getting them too much below the ground level, nor too high, so that they are much above it. The dressing knife should be kept very sharp to give a clean cut, as in all pruning." While this advice is good for England, Europe and eastern United States, on the coast we prefer to grub early, owing to the fact that we have not the extreme cold or heat that is liable to injure the young shoot.

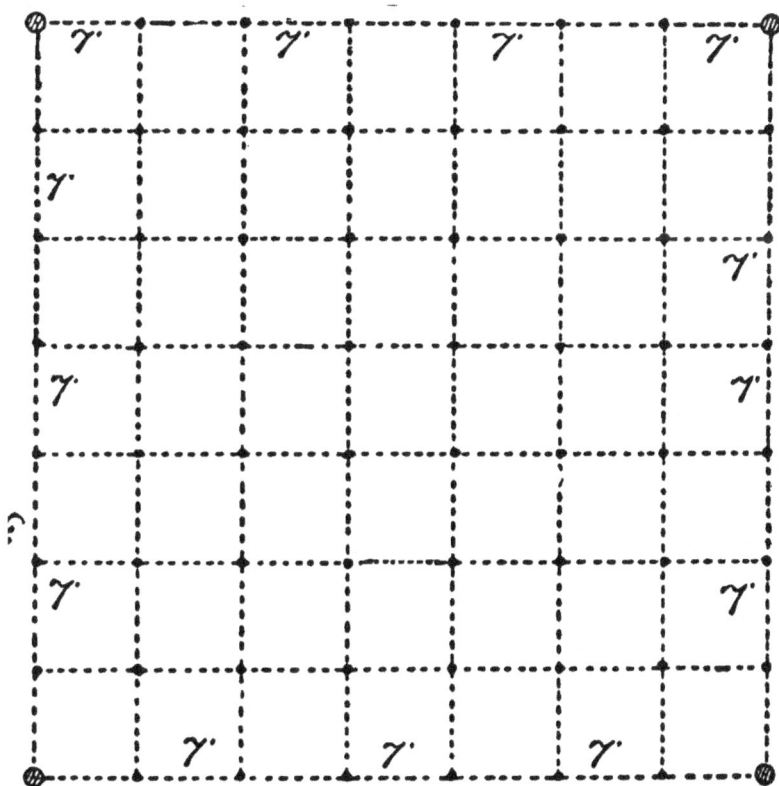

FIG. 43. YARD PEGGED OUT FOR PLANTING.

Perfect culture, from grubbing until the crop is laid by, is so important that we cite the various methods in vogue in all sections.

Washington.—Says Hart: "As soon as the shoots appear and are up about 12 inches, commence to train them by tying loosely with hop twine to the poles, nipping off all but two vines (some growers prefer only one). If you find some roots have not sprouted, examine and replace with others at once. This training must be continued until the vines have attained a height of some six feet, and even then, some of them may want training to prevent them breaking off. If the short pole system is in operation, you must now twine the yard, immediately after the second training, when vines are about four feet high. Long poles are not twined. Immediately after the first training, take a steady, gentle horse, with an eight-inch plow (this is large enough for the first time), and run a furrow on each side of the hills about one foot from the poles, throwing the soil from the hills, and then repeat on the lines of poles across on each side.

"This done, then by hand and a pronged hoe, work around the square untouched by the plowing at the foot of each pole, being careful to use a tool which will not hurt the root. This will prevent weeds growing around the vine. Next, take a steady, reliable team, and with a spring-tooth harrow cultivate between the rows, both ways, and thus thoroughly loosen the soil. This should be done twice during the growth of the vine to the top of the pole or peg. Now, take your team and plow up to the hills on each side and both ways and then follow both ways with a drag-tooth harrow. By this time your vines will be at the top of the short poles or to the peg of the long poles, and it is near time to commence spraying. When the hops are in the burr and forming, the yards should be gone over twice with the spring-tooth harrow and both ways to keep down the weeds."

Oregon.—Wolcott thus summarizes his own and the general practice: "Cultivation consists of first plowing the yard early in the spring with two horses and a turning plow, throwing the dirt away from the hill, then level down with either a cultivator or harrow and then cross-plow the same and level down again. After this, the yard should be gone over every two or three weeks with either a good cultivator or heavy disk harrow until about June 20, when all cultivation should cease, as cultivation after that date destroys the small feeders from the roots, which commence to shoot out near the surface and fill the space between the hills. Destroying these will cause the hops to take

86

another start and make them late in ripening. After all cultivation is done, the ground should be gone over each way with a clod masher or smoother, made the right width to go between the rows without damaging the vines. This levels and firms the soil and prevents evaporation during the long dry spell of July and August. The hop hills should be hoed as often as is necessary to keep down the weeds, and if none is permitted to go to seed for a few years, this will become a very small task."

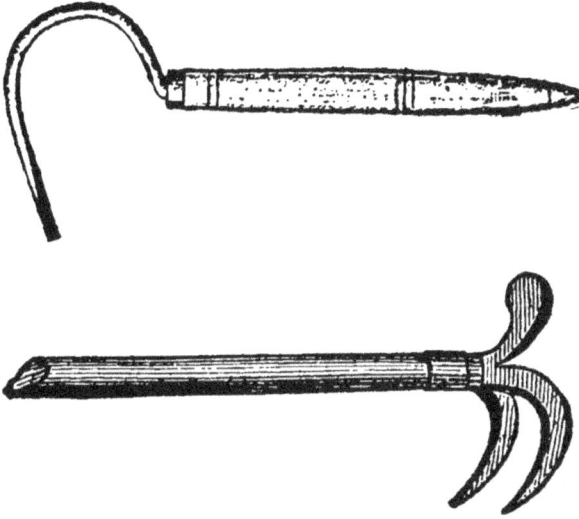

FIG. 44. AMERICAN GRUB HOOKS.

California methods are very similar. From three to five cultivations are given, according to condition of soil and weeds. Every effort is made to keep the soil open to a depth of four or six inches, and absolutely free of weeds. Hand hoe about the hills, to kill weeds and lighten soil not reached by the cultivator. Poled plants are usually hilled, but for the stake and trellis system level culture is generally preferred. See next chapter.

New York.—Clark puts the best practice in a nutshell: "As soon as the poles or stakes are set, start the cultivators, three or four times in a row, both ways, and keep going over the yard every week until within about two weeks of picking. Whatever may be neglected, don't fail to cultivate, cultivate, cultivate, as that loosens the soil, admits sun and air, releases the plant food, keeps down the weeds and advances and increases the crop very materially. Late cultivation also helps to bring the hops out of burr.

"About June 15-20, apply a good, large handful of phosphate, or fine ground bone, directly to the hill, among the vines, provided the vines are good strong ones; but if they are small or weak, place the phosphate a little way from the crown, because it might burn and injure weak, tender vines. Then hill in thoroughly, and the phosphate being applied to the crown, where the grubs work, together with the large, deep hilling, will help to drive the grubs out. As soon as the vines are well up the poles, say about four feet, cut off the surplus shoots, as they will begin to sap and weaken the hill. Then dress the hills out nicely with a four-tined fork, clean out all weeds and cover up any white hop sprouts that may be exposed, so as to protect them from the sun. If you have kept the cultivator going three or four times in a row every week up to June 15-20, you have the ground in a fine, clean, mellow condition, ready to fit up for billing. To do this thoroughly, run a large horse hoe through the center of the rows both ways, apply your phosphate or bone as mentioned before, and then hill in thoroughly and well up around the crown. If any extra sprouts have started out since the sprouts were cut off, hill them all in thoroughly, as they will help to keep the hill moist and will not bleed and weaken the hill, as they would if cut off. Have all vines kept well trained up on the poles or strings and keep the cultivators going every week, being careful not to dig into the hills just formed.

"About July 15, cultivate and horse-hoe thoroughly, the same as the first hilling. Then apply to the hill, among the vines, a good, large handful of unleached ashes and hill in well. Cover up all weeds and fill up all holes that have been dug out by skunks. About this time the hops will be in the burr. Sometimes they hang there a long time before coming out into the hop, and sometimes they fail to come out at all, or else have small, knobby and inferior hops. Cultivating, horse-hoeing, and billing at this time of the year help to bring them out of the burr more quickly and advance the crop to maturity. If any storms blow down the poles, they must be set up again as soon as possible."

Abroad, the plow is seldom put into a hop field, but the soil is turned over by hand with a spading fork or spud (Fig. 45) in late fall or as soon as the ground can be worked in spring. As soon as the vines have been tied up, a two-horse cultivator is run quite deeply into the ground, followed by a more shallow cultivation by lighter, one-horse hoe. The latter is used frequently until mid-July. The hand hoe is used to keep down weeds about the hills, and the soil about the hills not touched by the cultivator is worked once or twice with the Canterbury prong-hoe (Fig. 45). "Earthing, or putting earth over the stocks between the poles, is done by placing four or five shovelfuls of fine earth over them in June, to keep the bines in their places and to ensure a growth of roots for cuttings, or sets. It also stops the extraneous growth of bines from the stocks, which would exhaust them, and keeps them in their places." With slight modification, these methods prevail throughout Europe as well as England.

FIG. 45. FOREIGN HOP TOOLS.
1, Hessian pointed hoe; 2, English, Canterbury hoe; 3, English spading fork;
4, Bavarian broad hoe.

The practice of running the cultivator deeply in June, so as to break up the mass of fine rootlets from the hop roots, is adhered to by many careful growers, both in England and on the continent. The scientific reason for this practice has never been given, but probably is to be found in the theory that such root-cutting will force a new and fresh growth of rootlets, thus enabling the plant to feed more freely on the nutriment in the soil. Sturtevant applied this reasoning to his root-cutting culture of corn some 20 years ago, but in America, the theory finds few advocates in either the hop yard or corn field. The accepted plan is to give the hop rootlets a fine, mellow bed in which to flourish, with as little molestation as possible.

TYING UP THE VINES

As soon as the sprouts are up about three feet, tying up is in order. This is generally done by women, who take the best and most thrifty vines and wind them carefully around the pole, going with the sun, and tie them loosely with some soft material—matting bast, dried rushes, etc. Tie with a knot like Fig. 46, which will slip before it will cut the vine. The number of vines per pole varies from one to six, the larger number being where the training is on such a system as Fig. 50. Only the strongest vines should be tied up, the others being buried (not pulled up), though one or two may be left for reserve. The leaves on these buried vines will rot in a few days, making manure, and the vines will make cheaper food for the grub than those running up the pole. These buried vines throw out small roots, and help to feed the plant, and may furnish sets the next year. The yard must be looked over every few days to keep the vines well trained up and the heads must be kept free.

FIG. 46. TYING KNOT.

In the Fig. 3 system, when the vines get about eight or 10 inches above the nail, divide them and place two on each string and two up the pole, and then continue training until they get out of reach of the men standing on short stepladders. In the improved trellis system (Fig. 50), the vine has only to be given a few gentle turns around the strings and thereafter winds itself to the top without further assistance—a point vastly in favor of this system.

FIG. 47. A HOP GARDEN IN KENT. TWO POLES AND STRING.

A few days after the hops are laid by, that is, after the principal cultivation has ceased, the yard should be gone over and all leaves and arms should be cut off up to the height of a person's head. This will let the sun in underneath and will help in a great measure to keep down the ravages of the hop lice, as they first appear on the vines near the ground. In Washington, many growers turn sheep into their yards and let them eat off the leaves as high as they can reach. This is a very cheap method of cleaning the yard underneath, but where sheep cannot be had, a sharp knife must do the work. Flint cautions Californians not to trim off leaves unless foliage is very thick and ground very wet.

FIG. 48. PICKING HOPS GROWN ON STRINGS AND TRELLIS,
CALIFORNIA.

CHAPTER IX.
Methods of the Pleasanton Hop Company

[General statement of methods of cultivation, etc. employed by the Pleasanton Hop Co., at their yards in Alameda County, Cal.]

CLEARING—The vines are cut as soon after picking as practicable, generally during November when weather conditions and natural influences have killed the vines so as to prevent flow of sap from the root. The vines are cut at a point about 16 inches from the ground; the remaining portion of the old vine is cut at the crown of the root at pruning time in spring, while the portion that is cut as above mentioned in November, is immediately piled and burned, leaving the yard clean and ready for

(2) *Plowing,* which usually commences about February 1, with light two-horse single 10-inch plows. The earth is thrown by plow away from the roots toward center of row, as this method facilitates the work of

(3) *Grubbing* the roots, which is done by digging around the hills with a two-tined grape hoe, completely removing the earth from crown of root, care being taken not to bruise the main root, which is then ready for

(4) *Pruning*—This operation consists of removing with a sharp knife all the surplus small roots or "suckers" and cutting down the old wood of the previous year's growth, so as to make a new crown, from which start the vines intended for hop bearing.

Re-setting—Wherever, during the grubbing process, a hill is found to be defective or "missing," re-setting is done by planting three new roots or cuttings. These cuttings are about six to seven inches in length, and are planted so that the top of root is about level with the ground, and with the buds of the root pointing upward.

(5) *Covering or Hilling*—Immediately after pruning, as the work progresses, the roots are lightly covered with earth, using an ordinary hoe and making little mounds of earth, which serve to show the hop hills.

(6) *Cross-Plow*—After hilling, it is usual to cross-plow the hop yard, also away from the hills, leaving it in good shape for

(7) *Cultivation*—This has to be done at least twice or more, according to the season. The cultivators used here are *all iron* two-horse No. 3 McLean orchard cultivators (Fig. 49), having seven or nine standards, which can be used with either diamond shape or chisel teeth. A cut of the implement is shown herewith. The operation of cultivating also levels the land and returns the earth, which has been plowed away from the roots (see plowing). Now, to understand subsequent operations, it is necessary to describe the

FIG. 49. ORCHARD CULTIVATOR.

Trellis—The hop roots are planted seven feet apart, and at every sixth row a redwood pole 6x6 and 20 feet long is placed, being sunk two feet in the ground, and projecting, therefore, 18 feet above. The poles are thus 42 feet apart each way. All poles are dipped three feet in asphaltum tar at their planting end before being set. Heavy galvanized No. 2 wire is stretched across the tops of the poles in one direction (east and west), being fastened to the top of each pole with 2½ inch wire staples. *Directly over each row of hops* and resting upon this No. 2 wire, a smaller wire of No. 6 size is drawn (north and south) and is fastened to the larger wire wherever it crosses the latter. This fastening is done with small pieces of No. 18 wire. All interior posts are upright, while the outside rows of supporting wire poles incline at an angle of 30 degrees from the perpendicular.

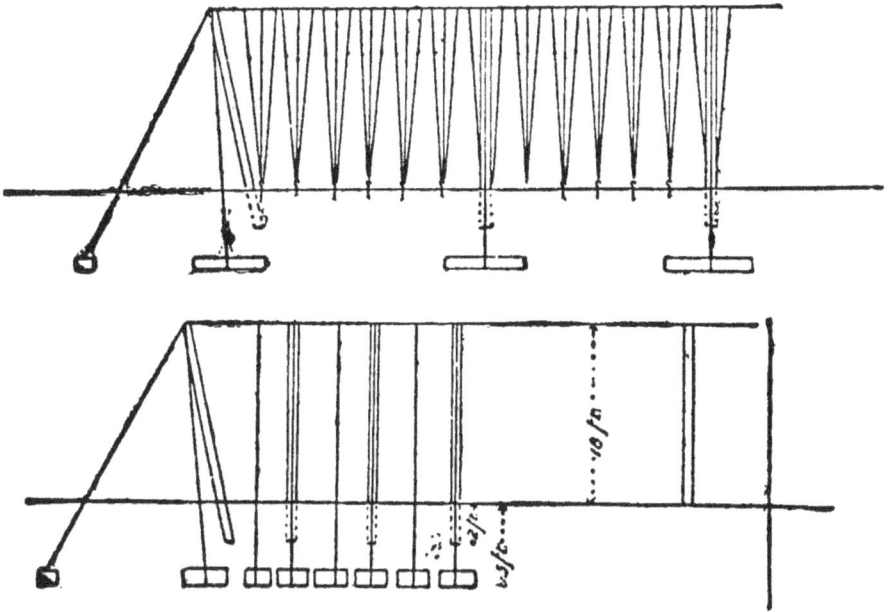

FIG. 50. PLEASANTON TRELLIS, SIDE AND END VIEWS

These supporting wires run in one direction only (east and west), and after winding around the head of the outside poles, at which point the wire is spliced, the wires pass at a downward angle of 50 degrees (or outward 40 degrees from the perpendicular) to the anchors. The supporting, or main-wire, *outside poles*, the same as all interior poles, are 42 feet apart.

The transverse, or trellis, wires, to which are attached the strings on which the wires climb, run north and south. The trellis anchor poles are set at an outward inclination of 20 degrees, and are placed only at the end of every alternating row of hills, making the distance between each pole 14 feet. The anchorage angles of these wires are the same as those of the supporting wires. The "alternate" rows of hills have no poles, the wires simply running over the end main or supporting wire to the ground and their anchorages, at the same angle as in the other cases. The supporting wires are re-inforced at the anchorages by six-strand, five-eighths-inch wire cables, spliced around the head of the outside anchor poles and running to the anchors, together with the supporting wire.

FIG. 51. STARTING OUT TO "STRING" A WIRE TRELLIS.

The trellis wires that run to poles are re-inforced by a No. 4 wire, joined to the poles as above, and running with the trellis wire to their anchorages. The alternate trellis wires above referred to as not having any poles, have no re-inforcement at their anchorages. All anchors are 6x6 redwood, are four feet long and are buried five feet in the ground (four feet deep we believe to be ample). The trellis should be erected in blocks of not to exceed 50 acres, and no stretch of wire should exceed 1,500 feet. It is even preferable to lessen this distance, and that anchorages be not over 1,000 to 1,100 feet apart, in large yards, so that they are in squares of 25 acres. This caution is given because experience has shown that where, from any cause or accident, the trellis poles collapse, or main wires break, the entire block within such anchorage is almost certain to go down. Thus it will be seen that the smaller the blocks, the greater the security; also the shorter the stretches of wire, the less the weight, and therefore the less liability to accident.

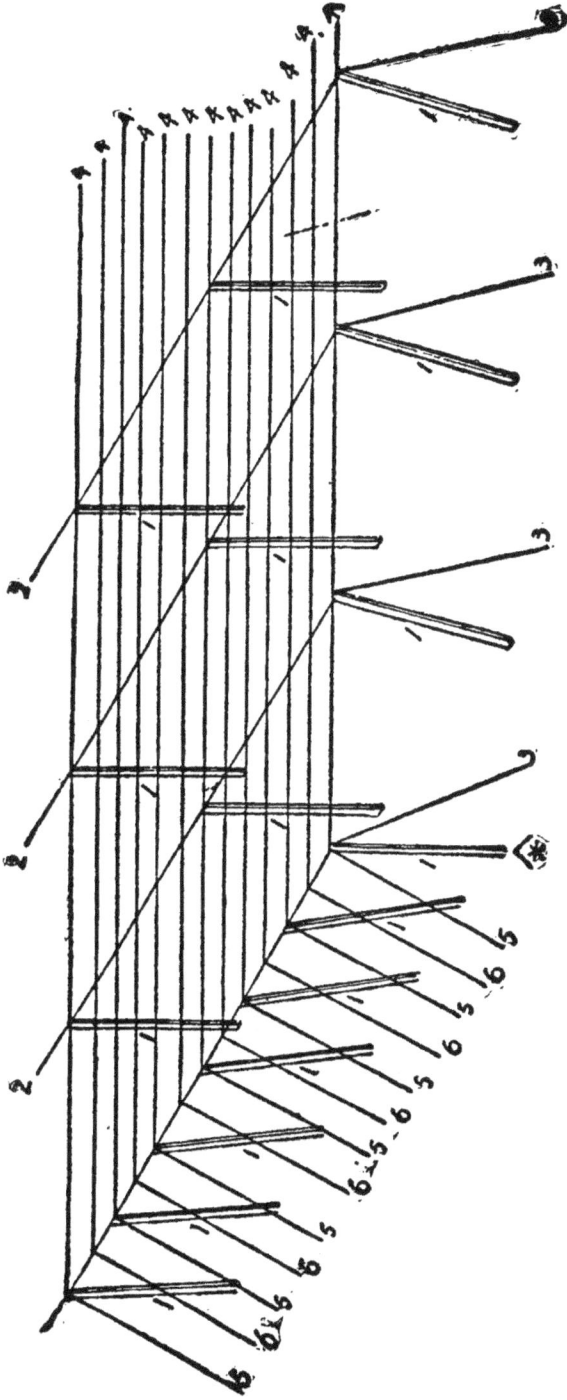

FIG. 52. TRELLIS SYSTEM USED AT PLEASANTON.

(8) *Stringing* begins about April 1. The method employed here is to tie three strings at the overhead trellis wire. The middle string is run perpendicularly from hill to wire, while the two outside strings run from hill to points about 20 inches on either side of the middle string. The strings are tied to the trellis wire first, and then all three are tied to a loop, in a wire stake about 15 inches long, which is shoved in the hill alongside the root.

(9) *Training*—About May 10, when the new hop vines are about two feet long, so that selection of the strongest (the hardiest, not the thickest) shoots can be made, one vine is trained upon each string. Care must be taken to train them as evenly as possible, from left to right; that is, following the sun.

(10) *Tuckering*—The surplus vines that are not used for hop bearing are pulled out. This operation is also necessary, as new shoots appear during the growth of the vine. Included in this process, it is usual to remove the lower arms or lateral growths of the vines on the strings, to a height of about three feet from the ground. Tuckering is done in order to throw the strength of the root into the main vine and arms.

Culture—During the above described process, the ground is well worked and cultivated with one-horse shovel plows (Fig. 53).

(11) *Hilling tip*—The final work of cultivation, about the first of July, is to plow a deep furrow each side of the vines, throwing the earth towards the roots, thus "hilling up" the roots, as in cultivating corn, etc. This is done with an ordinary one-horse plow.

FIG. 53. ONE-HORSE SHOVEL PLOW.

(12) *Clearing Wires*—When the vines are pulled down for picking, the string breaks close to the wire, thus leaving small pieces of the twine attached to the trellis wires. These become saturated with water from rain and dew and hold moisture, so that rust forms and weakens the wire at these points. For this reason, it is advisable to remove these bits of cotton by burning them with torches attached to long poles, as soon after harvest as possible. The gang that cleans the yard in November should perform this work, but when this is not practicable, these ends of twine should be cut or scraped when tying new strings to the wire in spring.

Setting Out Yard—As indicated under head of trellis, the main or support wires should always run east and west when possible, while the trellis or training wires should conformably run north and south, as this gives better sun exposure to the growing vines and hops.

FIG. 54. INDIAN HOP PICKERS AT DINNER, CALIFORNIA.

CHAPTER X.
PESTS OF THE HOP CROP

[The matter on insects affecting the hop plant, up to Page 125, was written for this work by L. O. Howard, Ph.D., Entomologist United States Department of Agriculture.]

ASIDE from the damage done by the hop plant louse in occasional seasons, and a rather infrequent period of abundance of the so-called "hop grub," the hop crop in the United States does not suffer seriously from the attacks of insects. The abundance of plant lice in general, with this species as with others, bears a direct relation to the weather, in that when the precipitation exceeds the normal, the lice are apt to be more abundant, whereas in dry seasons they almost entirely disappear. The same law also holds as to localities, and this doubtless accounts for the fact that the hop crop in England suffers more uniformly from lice than it does in central New York. The causes of the occasional abundance of the grub and of the other less important insects are not so readily determined.

Nearly all of the hop insects of the United States are native to this country and feed upon other allied plants of the family *Urticaceae.* The hop plant louse, however, is an exception and is of European origin, while there are one or two other European insects of some importance which feed upon this crop which may yet be introduced into the United States.

In this country there is no such thing as an annual drain upon the crop through the work of insects, although in an occasional season, as has just been hinted, the damage may be very great through the abundance of the lice. Such a season was that of 1886 throughout the hop belt of New York state. Some yards were completely ruined, while others lost from one-half to three-quarters of the crop. In Oregon and Washington, after the bad hop louse year of 1890, Professor Washburn estimated that one-twelfth of the crop of the states was ruined by lice and gave it a cash value of $365,000.

FIG. 55. SPRAYING OUTFIT, BRITISH COLUMBIA HOP YARD.

THE HOP PLANT LOUSE (*Phordon humuli*, Schrank)

In England, this insect has been a serious enemy of the hop crop for at least 200 years. The species is probably indigenous to that country, and has frequently been the cause of the trouble known to hop growers there as "black blight," the occurrence of which has increased apparently during the last 50 years. The crop in 1882, for example, was reduced from 459,333 cwts., to 114,832 cwts. The cost of picking the crop was reduced from £350,000 to about £150,000, so that not only did the owners of the plantations suffer, but the laborers who depended upon the hop picking were very considerable losers. The insect was probably introduced into the United States in the early part of the present century, and it is safe to say that it is one of the many species which have been brought to us upon nursery stock, since, as will be shown later, the insect hibernates in the egg state upon plum trees and is thus readily carried from one country to another, or from one part of the same country to another. It was not only brought to America from England in this way, but within recent years was carried from the east to the far west. As late as 1888 it was the boast of the hop growers of Washington and Oregon that they did not have to contend against the hop plant louse, but about 1888 or 1889 the insect made its appearance there, spread with the astonishing rapidity characteristic of plant lice, and, in 1890, accomplished the damage which we have noted in a previous paragraph.

The life history of this important insect has been fairly well understood in Europe for many years. It is remarkable from the fact it possesses a dual food-habit, living through the summer only upon the hop plant and passing the autumn, winter and early spring upon the plum. It is the first species of plant lice of which this peculiarity in life history was definitely proven, although it has since been shown to be common enough among species found in the summer time upon annual plants. It is strange that the discovery of this mode of life was not made earlier, since the necessity should have been obvious enough to anyone who might think about it. The necessity for this migration, however, is even more marked with the hop louse than with species feeding upon other annual plants, since not only does the hop vine die down in the fall, but the vines are generally pulled up and removed from the fields before they are killed by the heavy frosts of late autumn. The life history of the insect has accommodated itself wonderfully to this cultural practice and the lice acquire wings and leave the plant at just the proper time for the preservation of the species.

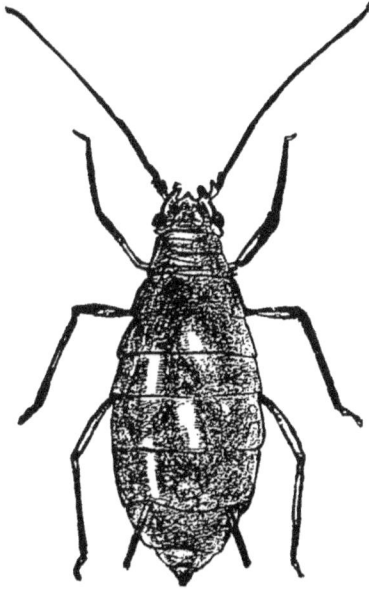

FIG. 56. HOP PLANT LOUSE.
True female.

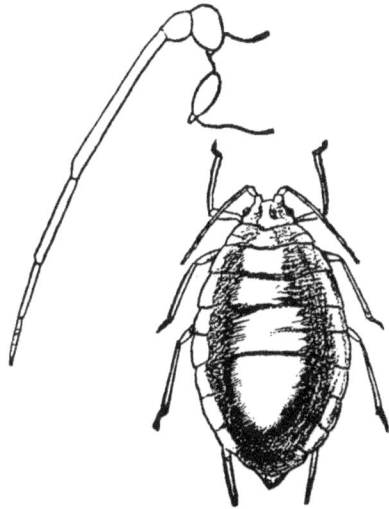

FIG. 57. HOP PLANT LOUSE
Stem mother.

Greatly enlarged. (From *Insect Life.*)

As before stated, the general relations of the insect to the two plants—plum and hop—were practically determined years ago in England, but it was only after the observations of 1887 made by the force of the Division of Entomology of the United States Department of Agriculture, and especially by Messrs. T. Pergande and W. B. Alwood, under the direction of the late Dr. C. V. Riley, that the full life round of the insect was known and the exact periods of development and of life upon either species of plant. These observations, which were carried out with the utmost care, and the results and methods of which form a model for similar work, have been recorded in Government publications, notably in the annual report of the United States Department of Agriculture for 1888, and in *Insect Life,* Volume I; also in Circular No. 2, Second Series of the Division of Entomology.

Life History—Briefly, it may be stated that the first plant lice in the spring hatch from winter eggs on the twigs of plum trees in the vicinity of hop yards. This first generation of lice is composed of wingless individuals which give birth to living young. These young settle upon the buds and young leaves of the plum tree, and after a few days give birth to other young. The second generation, like the first, is wingless, but the third acquires wings. There are no males among these lice, and the phenomenon of reproduction without the intervention of the male (termed parthenogenesis) is one of the wonderful features of insect life. By the time this winged generation makes its appearance, the hop plants in the yards have made a good start and the lice fly from the plum by common instinct to the nearest hop plants. Here they settle, immediately insert their beaks, and begin sucking up the sap of the plant, within a few hours giving birth to another generation of living young, which reach full growth without acquiring wings, just as did the first and second generations. There now ensue between the middle of June and the autumn, when the hop picking commences, from two to eight additional generations of these wingless virgin females.

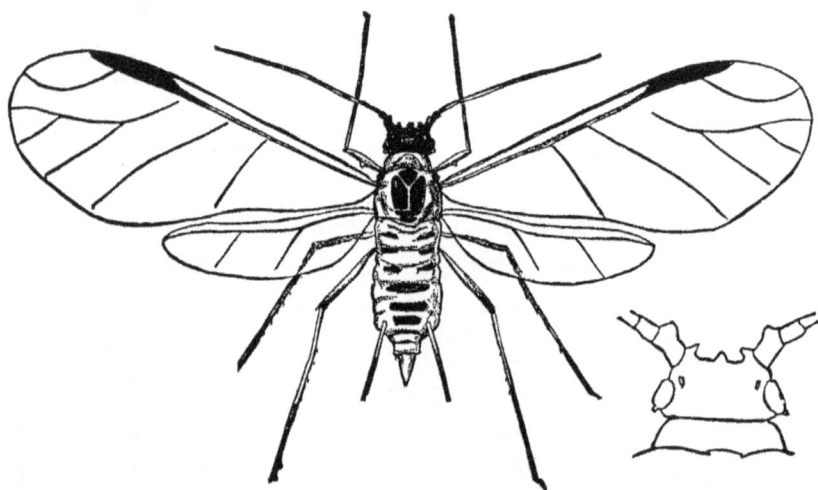

FIG. 58. HOP PLANT LOUSE.—(First Migrant).
Greatly enlarged (from *Insect Life*).

The rate at which they are produced is extraordinary. A female in the prime of life will give birth to several young each 24 hours. Each of these, in the course of eight days, becomes full grown and begins giving birth to young. Each female may live in the active, prolific stage for several weeks, so that a given individual may have living offspring to the fourth or even fifth generation before the end of her life. From this it results that from a comparatively small number of original migrants a large hop yard may be completely overrun with lice in a few weeks, under the most favorable circumstances. Were it not for the activity of the natural enemies of the lice, there would apparently be no hope of ever saving a crop. In September all the lice on the hop again acquire wings, whether they are of the fifth or the twelfth generation. We may have then ten wingless generations, and we always have two winged generations.

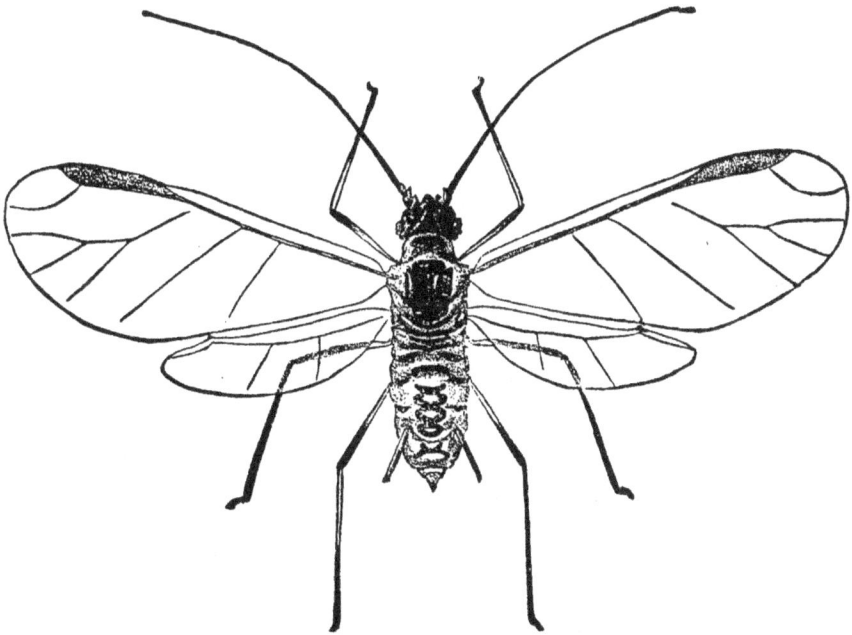

FIG. 59. HOP PLANT LOUSE.
Male greatly enlarged (from *Insect Life*).

The first individuals to acquire wings in the autumn are always females, and these leave the hop yards and fly back to the neighboring plum trees. The later individuals of this generation, and frequently, if not usually, all of the individuals of an additional generation on the hop, are true males, the male thus making its appearance for the first and only time in the annual life round of the species. By the time they have developed, however, the first issuing females will have settled upon the plum trees and will have given birth (parthenogenetically, as before) to a generation of wingless individuals which comprise the true females, not the virgin females as before, but the true females which must be fertilized by the males. So that, by the time the winged males have developed upon the hop crop and fly back to the plum, we have this generation of wingless, sexual, or true females awaiting them. Impregnation then takes place, the males die, and these wingless, sexual females give birth to the winter eggs, which are placed on the twigs of the plum, usually in crevices near the buds, and in this stage the insect passes the winter as before indicated. With the figures which are given of the different stages of the insect, no description is necessary and in fact all hop growers are familiar with the appearance of the green lice.

Natural Enemies—We have already mentioned the important part which the natural enemies of this insect play in its economy. In general, we may say that were it not for the natural enemies of plant lice and for occasional spells of extremely hot weather, all living vegetation would be destroyed by them. A great abundance of plant lice during a rainy spring is of common occurrence. Should the rains continue they increase beyond measure, but with the first stopping of the rain their natural enemies become active, reproduce with wonderful rapidity and destroy the lice by the wholesale. Then, too, when we have, as we occasionally do in late June or early July, a day or so when the temperature runs high up into the 90's, the lice may be killed off by the wholesale by the heat alone. The writer remembers a case in the city of Washington, where in a single day every plant louse of countless millions upon the box-elder shade trees was killed by a temperature of 101. At the same time hop plant lice which were under observation upon the grounds of the Department of Agriculture in the course of experimental work were also destroyed by the heat. It may be incidentally remarked that hot, sunny weather is prejudicial to the increase of this insect in the hop plantations of England. The natural enemies of the lice consist of the slimy maggots of several species of Syrphus flies, of the active and voracious larvae of the lace-winged flies, of the ladybird beetles, of the internal feeding larvae of an entire sub-family of parasitic flies known as Aphidiinae, and of several species of little parasitic hymenopterous insects, which, curiously enough, belong to the family

Cynipidae, most of the species of which are gall makers and plant feeders. Several of these insects are shown in accompanying figures. They need no detailed practical consideration, however, except that it should be pointed out that where any of these insects are very abundant, and the fact can readily be ascertained by a little close observation, remedial work may not be necessary. When any considerable proportion of the lice are found to be brown and swollen and nearly double the usual size, it is a sure indication that they contain Aphidiine parasites and that the little flies which will issue from these lice will be so abundant as to kill off the survivors without the necessity of remedial work. So, too, when ladybird beetles are very abundant upon the vines, it is safe to conclude that spraying with insecticide washes will not be necessary.

FIG. 60. HOP PLANT LOUSE AND EGGS.
Showing shriveled skin of female. Greatly enlarged. (From Insect Life).

Remedies—If the writer were a grower of hops, and had suffered from the attacks of this insect, his first step would be to locate every plum tree within a distance of half a mile from the hop yard. He would then, by hook, or by crook, secure the destruction of as many of these trees as possible, with the exception that he would leave two or three trees of moderate size among those nearest to the yard. These trees he would use as traps for the hop lice and every

107

spring, along toward the end of May, he would carefully examine the twigs, and, if lice were at all abundant, he would spray them thoroughly with a dilute kerosene soap emulsion, or with a resin wash. It is reasonably safe to say that if this course were or could be adopted by every grower of hops, comparative immunity of the crop from the attacks of these insects would be the result. There will be many cases, however, where there are so many plum trees near the hop field that such a course would be impossible on account of the value of the plum crop. This is apt to be the case especially in the extreme northwest, where the plum or prune crop is such a valuable one. In such cases recourse must be had to extensive spraying, preferably of the plum trees themselves in the early spring, since the lice are infinitely fewer in number at that time of the year, or of the hop crop itself, the earlier the better, after the lice make their appearance.

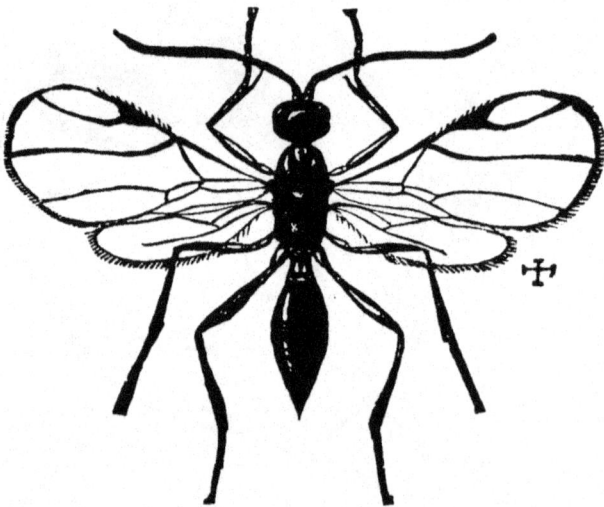

FIG. 61. APHIDINE PARASITE OF HOP PLANT LOUSE.
Greatly enlarged. (From U.S. Department of Agriculture.)

As to spraying materials, the extensive experiments which we carried on at Richfield Springs in the summer of 1887 show plainly the efficacy of the standard kerosene emulsion diluted with 15 parts of water and of a dilute soap wash made from homemade fish-oil soap. In Oregon and Washington, for some reason, the kerosene emulsion has not come into general use. As has been recently shown with such positiveness, in the case of the San Jose scale and the lime salt and sulphur wash, there is really a difference in the effect of the same insecticide wash on the Pacific and Atlantic coasts. Nevertheless, the decoction of quassia chips, which was so strongly recommended and so frequently used in Oregon and Washington in 1890 and 1892, fostered, as has been said, by the efforts of persons interested in the sale of the substance, when carefully tested by an agent of the United States Department of Agriculture in the field in Oregon in 1893, proved less effective than the kerosene emulsion and than the best of the fish-oil soaps, and it is probable that the disrepute into which the kerosene emulsion early fell was due to improper preparation and consequent destruction of foliage.

The standard kerosene soap emulsion formula is made as follows:

KEROSENE EMULSION

Kerosene	2 gals.
Whale-oil soap (or 1 qt soft soap)	½ lb.
Water	1 gal.

Dissolve the soap in boiling water and add the hot solution, away from the fire, to the kerosene. Agitate violently for five minutes by pumping the liquid back upon itself with a force pump until the mixture assumes the consistency of cream. In this condition it will keep indefinitely and should be diluted only as wanted for use. For plant lice and other soft-bodied insects, dilute the above to 15 or 20 gallons. For scale insects and beetles use seven to nine parts of water.

Fish-oil soap is made in the following way: Take potash lye, 1 pound; fish oil, 3 pints; soft water, 2 gallons. The lye is dissolved in the water, and when brought to the boiling point the oil is added. The batch is boiled for about two hours. Enough water is filled in to make up the evaporation by boiling, and the result will be about 25 pounds of soap, which, when cold, may be cut and handled in cakes. This is enough for 150 gallons of effective wash and will cost from 20 to 25 cents in Oregon.

Additional experiments were made in 1893 with resin wash and the results were very satisfactory. The formula used by the agent, Mr. Koebele, that year was as follows: One pound of caustic soda dissolved in two gallons of water and six pounds of broken resin, to be boiled with about three quarts of the resultant lye. After the resin is dissolved, the rest of the lye is to be added slowly, with water to make about eight gallons of the compound, which should be still further diluted with water before cooling. The resulting mixture should be clear and brown in color, and at this stage it is readily diluted with water. Improvements have been made since 1893 with the wash and the formula now recommended by the writer's office is as follows: Resin, 20 pounds; crude caustic soda (78 per cent.) 5 pounds; fish-oil, 2½ pints; water to make 100 gallons. Ordinary commercial resin is used, and the caustic soda is that put up for soap establishments in large 200-pound drums. Smaller quantities may be obtained at soap factories, or the granulated caustic soda (98 per cent.) used—3½ pounds of the latter being equivalent to five pounds of the former. Place these substances, with the oil, in a kettle with water to cover them to a depth of three or four inches. Boil about two hours, making occasional additions of water, or until the compound resembles very strong, black coffee. Dilute to one-third the final bulk with hot water, or with cold water added slowly over the fire, making a stock mixture to be diluted to the full amount as used. When sprayed the mixture should be perfectly fluid, without sediment, and should any appear in the stock mixture, reheating should be resorted to, and in fact the wash is preferably applied hot. These resin washes, it should be stated, are applicable only in regions where there are comparatively long rainless periods, since they are readily washed from the trees by rain. The standard wash now in use in the state of Washington consists of six pounds of quassia chips and five pounds whale-oil soap to 100 gallons water, and it is said that many growers get excellent results from this mixture.

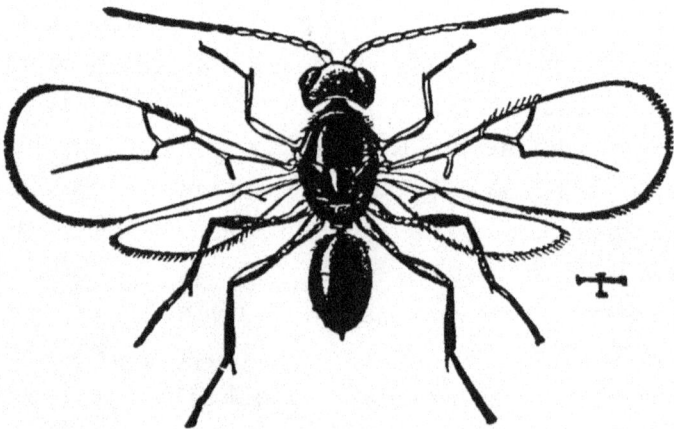

FIG. 62. CYNIPID PARASITE OF HOP PLANT LOUSE.
Greatly enlarged. (From U.S. Department of Agriculture.)

110

As to apparatus for the application of these insecticides, little need be said. So many excellent machines are on the market that the hop grower will have no difficulty in selecting one suited to his needs and the condition of his finances. Homemade machines consist simply of a barrel mounted upon a sled with a pump inserted in its top. Long ¼ inch 3-ply hose, bearing "cyclone" nozzles and supported by bamboo poles, afford easy means of reaching all parts of the plants.

Conditions in the Different Hop-Growing Regions—We have referred in our introductory paragraph to the fact that this insect brings about appreciable damage only occasionally. Thus, in the great hop-growing region of central New York there have been no fields totally destroyed by the louse since 1886. In 1891, in the early part of the season, there was a hop louse scare among growers and considerable damage was anticipated, but with the dry season in July the insects almost entirely disappeared. There have been more or less lice every year since 1886 and some fields have turned out a poor quality of hops in consequence of mold in the burr, caused indirectly by the lice and by damp, warm climatic conditions just before harvesting. It seems to be an accepted fact that those growers who are free from wild plum trees and have their yards on the upland usually have less mold than growers with yards along lake shores.

In Oregon, the damage was greater in 1890, two years after the introduction of the insect, than it has been since, with the possible exception of 1897. In 1891 there was less injury than in 1890; 1897 was a year of considerable damage, the extent varying from 2 per cent. of the crop in one yard to 95 per cent. in another. The average loss, the state entomologist (Prof. Cordley) states, was about 33 1-3 per cent. It was fully 50 per cent. in the neighborhood of Corvallis.

In Washington the damage has been more or less constant since 1890. I am informed by Prof. Piper, state entomologist, that some of the best hop growers in the state of Washington grow A 1 hops with but one spraying, using whale-oil soap and quassia chips, while others spray two or three times. In the Yakima valley, he states, the summer heat is so great that the louse succumbs to it, although it may be abundant early in the season, and it has not been necessary to spray in that region, which, by the way, is irrigated, sage-brush land. Nevertheless, on account of the expense of spraying in western Washington, Professor Piper is of the opinion that hop growing will never again become the industry that it was prior to the introduction of the louse. There is some prevalent opinion in Washington that the life history given in preceding sections will not hold for that part of the country. Growers claim that winged lice occur throughout the season and they do not believe that all the winter eggs are deposited on prune or plum trees. This statement seems extremely improbable to the writer, but it must be stated that no observations have been made in that part of the country which are at all comparable with the extremely careful ones carried on in New York in 1887. Nevertheless, Mr. Koebele, when in Oregon in 1893, was able to set at rest one of the local misapprehensions, which was to the effect that the hop louse occurs also on one of the mints. He sent specimens of the insect to the city of Washington, where, upon examination, it was found that although the resemblance was extremely close, the mint insect belonged to a different species of the same genus, Phorodon.

The hop plant louse made its first appearance in the Wisconsin hop district in 1867-68, and from that time on was more or less abundant every season, some years almost entirely destroying the crop, and in others causing only partial loss. Its attacks have practically ruined the hop industry of the state.

THE HOP GRUB OR HOP-PLANT BORER
(*Gortyna immanis*, Grt.)

This insect probably ranks second in importance among those which we shall mention, although of late years it has been vastly less destructive than the plant louse. It is a distinctive North American insect and is known as a hop pest only in the east. The moth has been found in the state of Washington, but the grub has not been reported to damage the hop yards in that state.

In 1882 the insect was brought to the attention of Professor Comstock, of Cornell University, who first learned its complete life history, and in 1883 it was investigated by Dr. J. B. Smith, then an agent of the United States Department of Agriculture. Since that time no reports of serious damage have been received. A prominent hop grower writes me from Richfield Springs, N. Y., under date of January 20, 1898, that the grub usually eats off some vines, but seldom does much damage. Skunks, he writes, are plentiful, and they dig the grubs out of the hop fields in the summer.

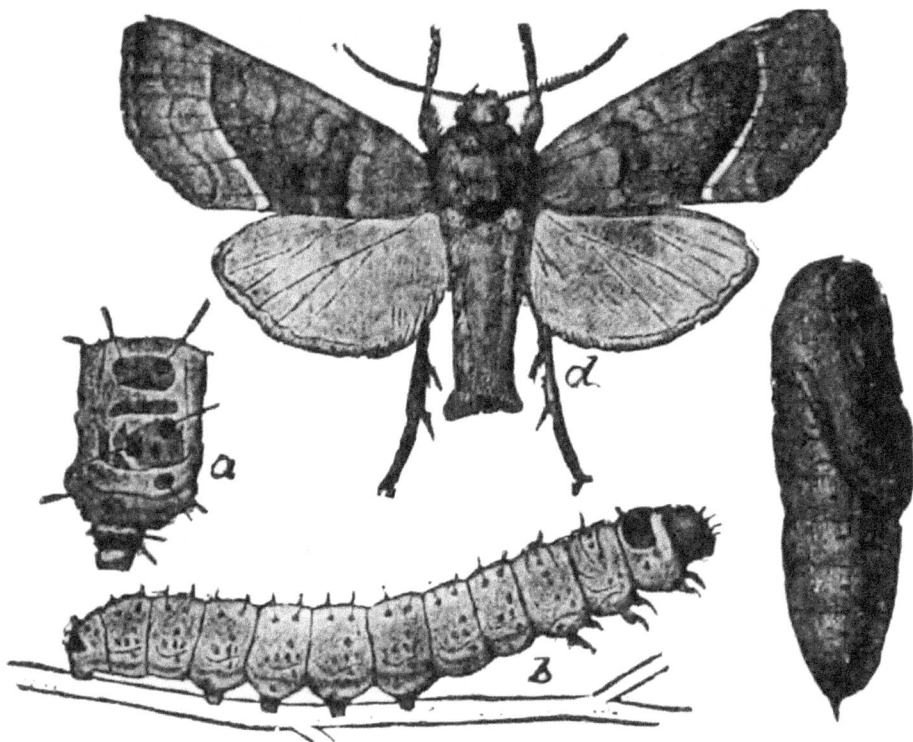

FIG. 63. HOP GRUB

a, Segment of larva; *b*, larva; *c*, pupa; *d*, adult. All natural size except *a*. (Author's illustration.)

The adult moth of the hop grub, shown at Fig. 63, lays its eggs in the early part of the season upon the young shoots of the plant. The young caterpillars, which are slender and greenish in color, spotted with black, bore into the vines just below the tip and remain at this point for some time. The head turns downward and stops growing. Such vines are called "muffle heads" or "stag vines," and sometimes "bullheads," by the growers, and the caterpillar inhabiting them is called the "tip worm." A little later the grub drops to the ground and enters the stem at the surface of the ground. It is then called the "collar worm." It changes to a dark, whitish color with black spots. About the end of July or the first of August it becomes full-grown, and transforms to pupa near the roots of the plant. The moths seem to issue in part in the fall and in part in the spring, and the insect, therefore, passes the winter in the moth state under rubbish and in fence cracks, as well as in the pupal state underground.

As to remedies, where the insects are really abundant, it is always desirable that the men engaged in tying the vines should pinch off affected tips and crush the worm. Many of them are easily destroyed in this way. Others, however, escape, drop to the ground and begin work at the crown. A generally adopted remedy at this time is high hilling and fertilizing, which induces the putting out of rootlets above the main root, enabling the vines to derive nourishment through this channel when the stem has been gnawed through. An experienced grower in Otsego county, N. Y., recommends that at the first hoeing the dirt be carefully worked away from the vines by the hoe, leaving them bare down to the bedroot. The weather toughens the lower part of the stem and renders it unattractive to the grub. Immediately after the hoeing, a handful of composite, consisting of equal parts of salt, quicklime and hen manure, mixed while slaking the lime and left standing for two weeks, should be placed about each vine root.

CATERPILLARS FEEDING UPON HOP LEAVES

Several different kinds of caterpillars feed on the leaves of the hop plant during the summer, but they are easily controlled and seldom do any especial damage. Certain of these species may be illustrated and briefly mentioned. All are readily destroyed by an arsenical spray. Should any one of these insects become sufficiently numerous to threaten damage, and any of them is at all times liable to sudden increase, the yards should be promptly sprayed with Paris green or London purple, in the proportion of one pound to 150 gallons of water, or with arsenate of lead in the proportion of two pounds to 100 gallons of water.

THE HOP VINE SNOUT-MOTH (*Hypena humuli*, Harr.)

In 1856 Dr. Fitch, writing of this insect, considered it to be the most universal and formidable of the hop insects, making its appearance suddenly, and sometimes in a few days completely riddling and destroying the leaves of whole fields. The rather slender, green caterpillars make their appearance in the latter part of May, feed upon the substance of the leaves until full-grown, and then form thin, imperfect, silken cocoons within a folded leaf or in a crevice or sheltered spot, transforming to chrysalids and issuing as moths three weeks later. There are two annual generations, the second brood of caterpillars being found upon the vines in August. The insect hibernates in the moth stage.

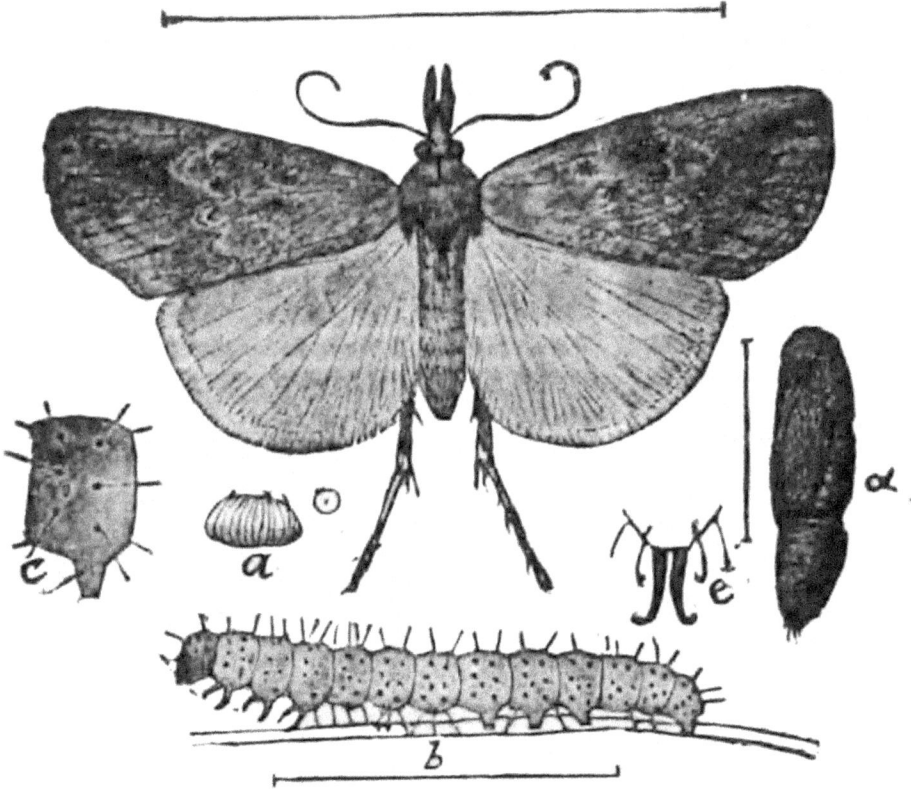

FIG. 64. HOP VINE SNOUT-MOTH.
a, Egg; *b*, larva; *d*, Pupa; *f*, moth. All enlarged; natural size indicated by dots and hair lines. (Author's illustration.)

HOP MERCHANTS (*Polygonia interrogationis*, Godart, and *Polygonia comma*, Harr.)

These are common, widespread, and, in the adult stage, handsome butterflies, occurring in most parts of the eastern United States, and in the caterpillar stage feeding not only upon the hop, but also upon the elm and several other closely allied plants. They have derived their name of "hop

115

merchants" through the gold and silver markings upon the chrysalids, which occasionally, probably through parasitism, become suffused and give a general golden or silvery tinge to the chrysalids. As I have shown in another publication, an interesting superstition is more or less laughingly held among New York hop growers, to the effect that when the golden spots are plentiful, the crop will be good and the price high, while, if the silvery cocoons are more abundant, the price will be low. Both of these insects are double-brooded in hop-growing regions, and they are shown in their different stages in the accompanying figures. The spiny caterpillars are readily recognized, and feed without concealment upon the upper or under surface of the leaf. They are frequently extensively parasitized in the caterpillar stage, as well as in the egg stage, by minute hymenopterous parasites, a fact which accounts in large measure for the slight damage done by these insects under ordinary circumstances.

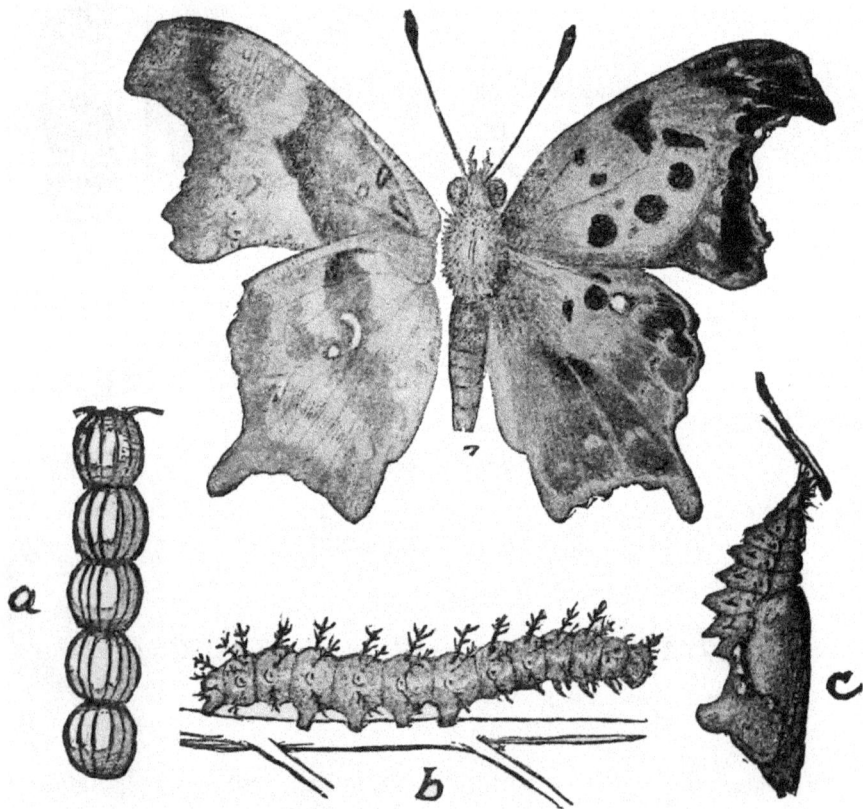

FIG. 65. INTERROGATION BUTTERFLY.
a, Eggs; b, larva; c, chrysalis; d, adult. All natural size except a, which is enlarged.
(Author's illustration.)

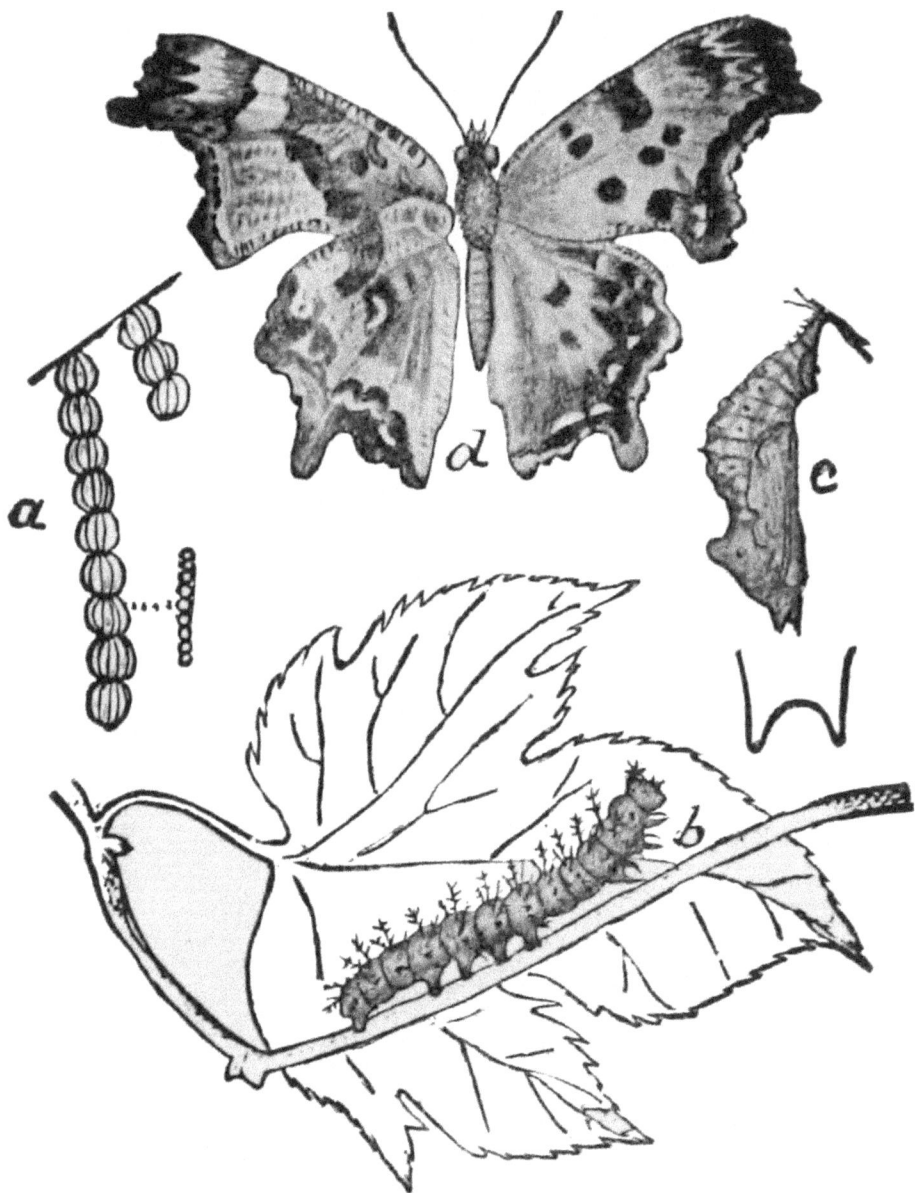

FIG. 66. COMMA BUTTERFLY.
a, Eggs; *b*, larva; *c*, chrysalis; *d*, adult. All natural size except *a*, which is enlarged.
(Author's illustration.)

THE ZEBRA CATERPILLAR (*Mamestra picta*, Harr.)

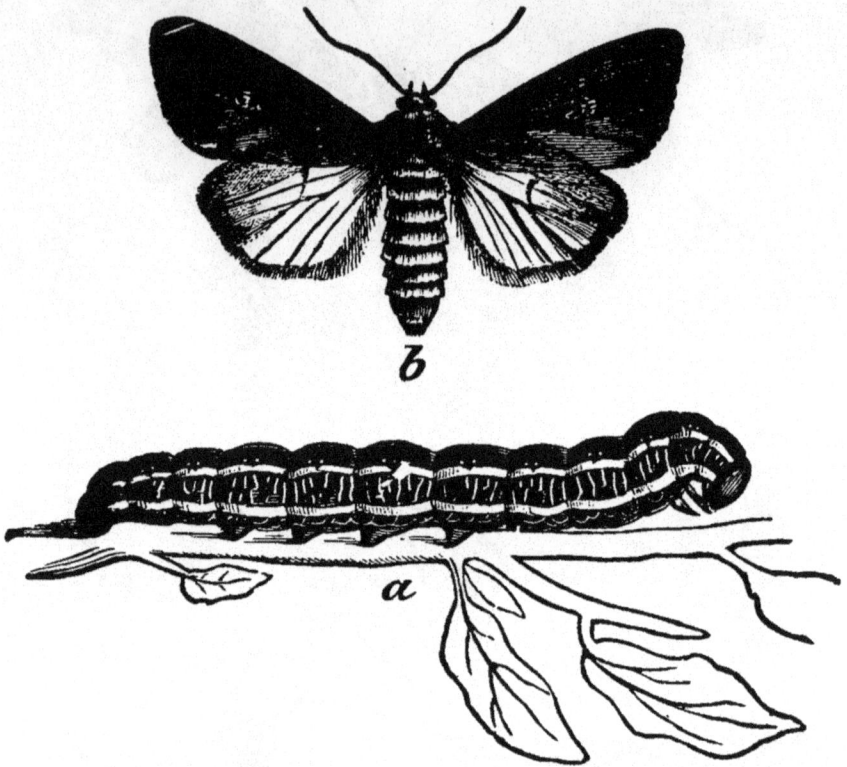

FIG. 67. ZEBRA CATERPILLAR.
a, Larva; *b*, adult. Natural size. (After Riley.)

This well-known and polyphagic insect is found frequently upon hops. It occurs from Canada south to Virginia, and west to Nebraska, and has evidently of late years been carried into California. It feeds upon blackberry, poke weed, lamb's quarter, goose foot, worm seed, cabbage, aster, honeysuckle, white berry, mignonette, asparagus, ruta-baga, beet, cauliflower, spinach, bean, pea and celery, and is thus a common garden pest. The eggs of this insect are deposited on the lower sides of the leaves in clusters of from 250 to 300. The young caterpillars, at first almost black, but afterwards pale green in color, feed together in bands on the undersides of the leaves. When they reach the third stage, they begin to scatter, and thereafter feed singly, assuming a velvety black color, with two narrow yellow lines down the sides, between which are

118

numerous transverse irregular finer, yellow lines. When full-grown, they burrow into the ground and change to pupae in about two days. The insects pass the winter in the pupal stage, the moths issuing in May and June. The young caterpillars are found from the first week in June to the first week in July, and reach their full growth in about four weeks. A second brood of moths in more southern localities appears during the early part of July.

THE COMMON WOOLLY BEAR CATERPILLAR
(*Spilosoma virginica*, Fab.)

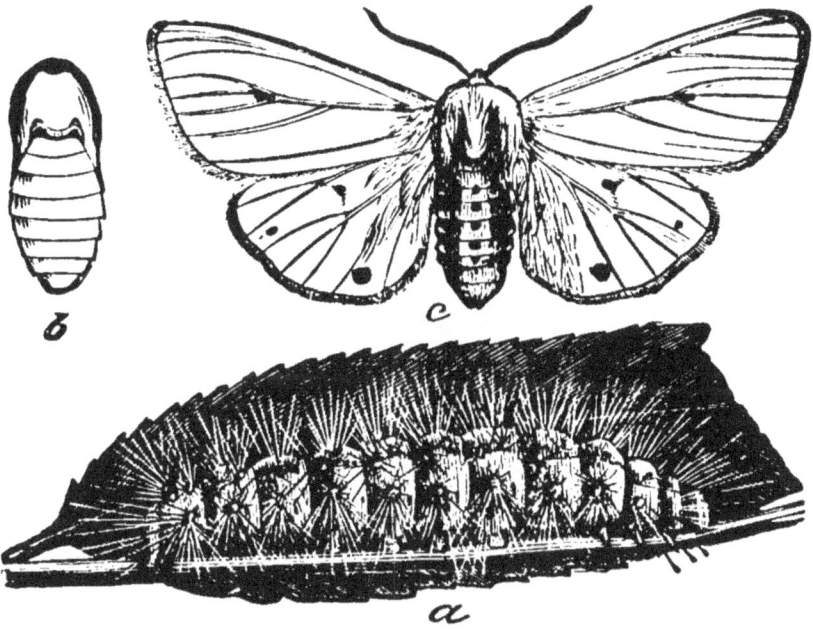

FIG. 68. WOLLY BEAR CATERPILLAR.
a, Larva; *b*, pupa; *c*, adult. Natural size. (After Riley.)

This is another common, widespread species with many food plants, which is quite often found in the hop yards, feeding upon the weeds, as well as upon hop vines. The caterpillars when full-grown are very voracious and will devour an entire leaf in an incredibly short time. They are not frequently seen upon the plant, since they drop readily when disturbed and remain quiet for a few minutes, but they are quick travellers when in motion. The eggs are laid on the lower sides of the leaves, in batches of from 50 to 100 or more. The full-grown larva is an inch and a half in length and very variable in color. It is covered with stiff hairs, which are sometimes white, intermixed with a few yellow or brown ones, or they are yellow, red, brown, or almost black, sometimes darkest at both ends, or all colors mixed. The cocoon is mostly composed of hairs of the caterpillar, and is spun in any suitable sheltered position. There is apparently but one annual generation, and the insect hibernates both in the caterpillar stage and in the pupal stage in its cocoon. The figure which we give represents perhaps the commonest variety of the caterpillar.

THE SADDLE-BACK CATERPILLAR
(*Empretia stimulea*, Clem.)

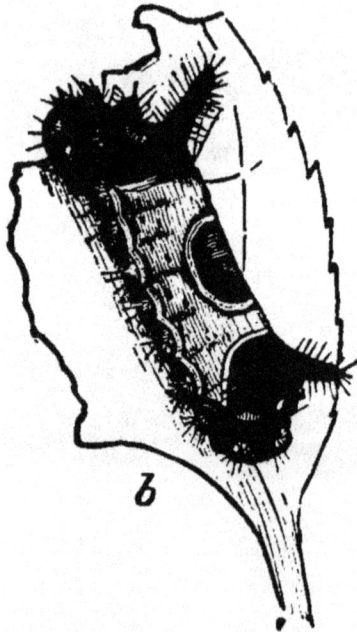

FIG. 69. SADDLE-BACK CATERPILLAR.
Natural size. (After Riley.)

This insect is another of the general feeders, and will probably not play an important part as a hop insect, for the reason that it is a normal denizen of regions too far south for the successful commercial cultivation of the hop. In fact, the only hop-growing region where it has ever been found is in southern Wisconsin, and, as has been shown, the culture of hops has been largely abandoned in that state. It occurs, however, commonly upon the hop vines grown in the dooryards throughout the southern and mid-western states, and will readily be recognized from the accompanying figure. It is one of the stinging or urticating caterpillars, and its spines coming in contact with a delicate skin have very much the effect of one of the nettle plants. The insect over-winters in the pupal state within its cocoon, and there are two or more generations each year.

OTHER CATERPILLARS

Descriptions of the remaining leaf-feeding caterpillars will hardly be necessary in this connection. The species found most commonly upon the hop are as follows: *Thecla humuli*, Harr.; *Ctenucha virginica*, Charp.; *Acronycta brumosa*, Guen.; *Acronycta americana*, Harr., *Orgyia antiqua*, L.; *Halisidota caryae*, Harr.; *Halisidota tessellata*, S. & A.; *Plusia precationis*, Guen.; *Leucarctia acraea*, Dru.; *Hypena scabra*, Fab., and *Hyphantria cunea*, Dru.

LEAF HOPPERS WHICH AFFECT THE HOP

Several species of the little insects properly called leaf hoppers, but which vine growers have become used to calling "thrips," occur upon the hop plant, and in dry seasons sometimes cause the leaves to turn brown and wilt, thus doing about the same character of damage in dry weather which the hop plant louse does in damp weather. The most serious case which has been brought to our attention was in 1891, when specimens of the species here figured, namely, *Tettigonia confluenta*, Say, were received from Puyallup, Wash., in August, with the statement that they were very numerous upon the blossoms or cones, and were injuring their quality to some extent. Further reports of damage of the same nature have not since been received, but it is an insect which hop growers of the northwest should know and should guard against. In the east, the most abundant of the leaf hoppers found in the hop yards is *Typhlocyba rosae*, numbers of which were found in the yards at Richfield

121

Springs, N. Y., in June, 1887, causing more or less damage to the foliage. Another species, more closely related to the one found in the state of Washington, was collected in numbers on the hop vines at Waterville, N. Y., in July, 1883, by Dr. Smith. It is a handsome species of the genus *Typhlocyba*, and is of a yellowish-green color. Dr. Smith found that yards badly affected with lice had none of these hoppers, while in yards in which the lice were absent, the hoppers were more numerous.

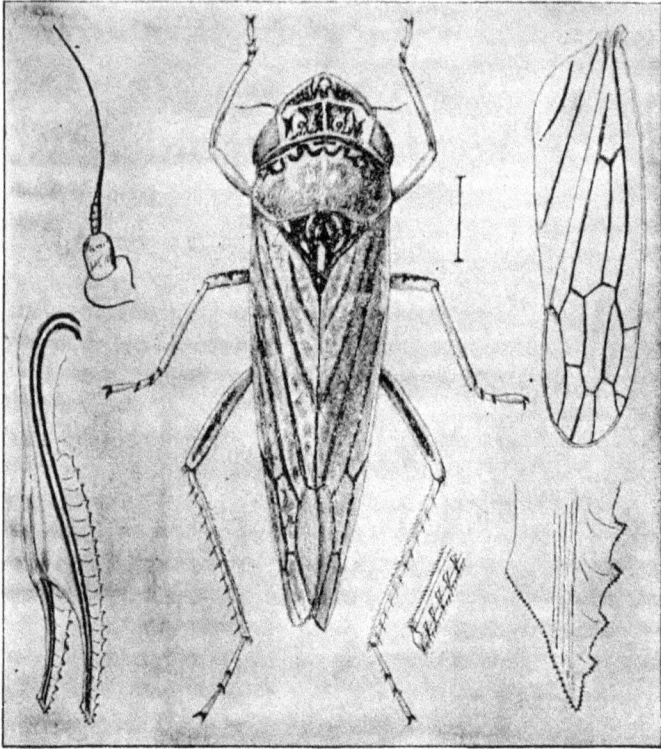

FIG. 70. HOP VINE LEAF HOPPER.
Tettigonia confluenta, with enlarged structural details. Enlarged (original).

Nearly all of these leaf hoppers over-winter in the adult condition, under leaves and rubbish at the surface of the ground. A hop yard, therefore, which is thoroughly cleaned up in the autumn, and all leaves and rubbish burned, will generally be free from this insect. Where they are very abundant in the summer time, there are two remedies which may be adopted. The great activity of these insects under ordinary circumstances makes spraying ineffective, but during the early morning or late in the evening—especially on a cool, moist day—they are more torpid and can then be struck by a spray of kerosene emulsion. A method which has been adopted in New York vineyards and also to some extent in California vineyards, is to make a light shield of a lath frame, with cloth stretched over it, and this, when saturated with kerosene or painted with tar, is carried through the field to the leeward of the vines, the vines being stirred on the other side. The hoppers fly against the kerosened or tarred surface, and are thus destroyed in large numbers.

BEETLES FEEDING ON HOP LEAVES

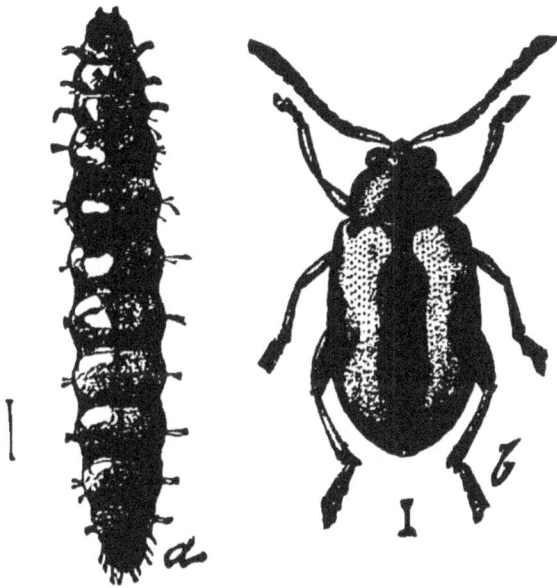

FIG. 71. STRIPED FLEA-BEETLE.
a, Larva; *b*, adult. Enlarged. (From U.S. Department of Agriculture.)

Several species of leaf beetles are frequently found in the hop yards, and gnaw holes in the leaves, thus disfiguring them, but seldom injuring the plant. Among these are the red-headed flea-beetle (*Systena frontalis*), the striped flea-beetle (*Phyllotreta vittata*), the punctured flea-beetle (*Psylliodes punctulata*), and the twelve-spotted leaf beetle (*Diabrotica 12-punctata*). These species were found by Dr. Smith at Waterville. Mr. Pergande, at Richfield Springs, collected *Phyllotreta vittata; Crepidodera helxines*, the common willow flea-beetle; *Epitrix cumumeris*, the potato flea-beetle; *Diabolia borealis*, the

123

common plantain leaf miner; *Psylliodes punctulata,* a common and widespread species. None of these insects is peculiar to the hop plant. The common striped flea-beetle is shown in the accompanying figure.

THE SO-CALLED "RED SPIDER," OR SPINNING MITE

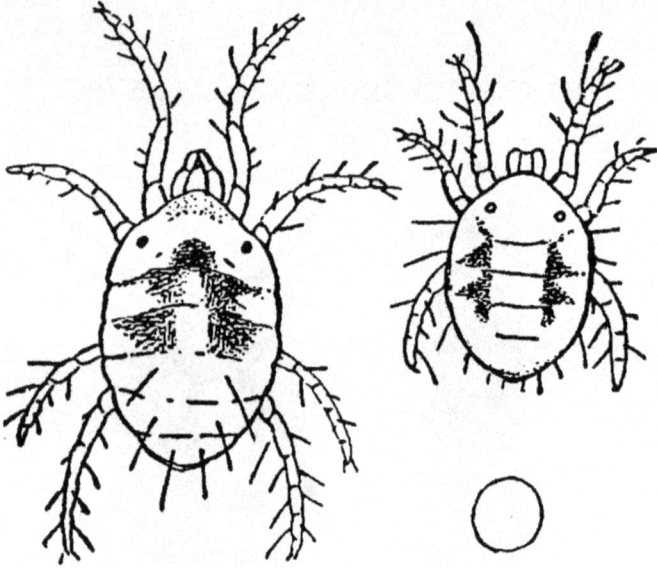

FIG. 72. "RED SPIDER," OR SPINNING MITE.
Female, male and egg-greatly enlarged. (Redrawn from Journal of the Board of Agriculture of England for December, 1897.)

Hop fields in England have occasionally suffered to a considerable extent from the damage done to the foliage by what appears to be the common red spider of our American greenhouses (*Tetranychus telarius*), although English writers have found sufficient difference between the spinning mite found in the hop fields and the ordinary form to establish a new variety which they call *T. telarius* var. *humuli.* In 1868, and again in 1893, this little mite did much mischief in many hop yards. The leaves fell off, the burr or blossom was damaged, and in some instances the plants were completely shrivelled up. In 1897, again serious injury was threatened, but a succession of showers and a fall of temperature fortunately checked multiplication. The first indication of the presence of these mites is the yellowing of the lower leaves of the plant,

124

and when examined carefully upon the under surface, thick, silken webs will be seen spreading from rib to rib, under which the mites live, actively sucking the juices of the leaf. The remedies adopted in England are very sensible, and consist, principally, of a heavy spray of soap and water and sulphide of potassium. Sulphur in any form is a specific against mites, and a spray of kerosene soap emulsion, to which a small quantity of flowers of sulphur has been added, is generally effective.

I am not prepared to say that this same mite is found in American hop fields, but in September, 1887, Mr. Pergande found at Waterville, N. Y., a species closely related to the common red spider of green houses, which occurred in large numbers on the lower side of many leaves of the hop plants, doing considerable damage to the foliage and covering themselves with a web just as the spinning mite of the hop fields of England is reported to do. It bore a strong superficial resemblance to the common so-called red spider, but had six-jointed legs instead of seven-jointed legs.

Professor Osborn, in Wisconsin, in September, 1887, found what he took to be the true *Tetranychus telarius* in almost every hop yard visited, and in some so plentiful as to cause conspicuous injury to the leaves. It should be stated that in Wisconsin that summer the hop plant louse seemed to be entirely absent. He found eggs, young mites and full-grown mites abundantly under the very delicate web spun over the under surface of the leaf, the upper surface indicating their presence by rusty patches and a red or yellow discoloration. No attack was made on the burr, so that the damage consisted simply in loss of vitality to the plant. The growers generally did not consider the mite as of any importance. Prof. Osborn has suggested the obvious remedy of burning the plants as soon as they become dry enough to burn after picking. Thus, there is a possibility that the European mite already occurs in this country, and that trouble may ensue in exceptional seasons.

PRACTICAL DIRECTIONS FOR SPRAYING

Such poor results from spraying have been reported that in addition to Dr. Howard's very complete and scientific exposition of the subject, we add some directions and experiences from practical growers who have successfully applied the foregoing methods. Writing for Washington state, Hart says:

"The mixture most esteemed here is quassia chips and whaleoil soap. For each acre to be sprayed, soak in cold water the first time 10 pounds of quassia chips in 25 gallons of soft water; the second time you will boil them for two hours. Also boil five pounds of whaleoil soap in 25 gallons of soft water until the soap is thoroughly dissolved; then strain the same, mixing them alternately into a clean barrel, a bucket at a time, and stirring together.

"Place your barrel, which should have a force pump attached thereto, upon a good sled, with one man to pump and drive (a steady horse being necessary), and two men to spray, one on each side. Each man sprays two rows of vines, making four rows in all sprayed at one time. Use a fine rose nozzle, being especially careful and particular to spray the underside of the leaves. All the men should be clothed in oil or green coats and hats, to protect them from the spray, for in a wind it is almost impossible to do good work. There should not be less than three sprayings, four is safer, the last one just as the hops are forming, and the liquor that time may be slightly reduced in strength so as not to injure the hop. In the short pole system, one objection is the difficulty of getting through without severe scratchings, and the team often is entangled in the vines crossing over the twine above their heads."

For Oregon, Walcott writes:

"In the future, we cannot count on a crop of good quality without spraying. There are many methods and formulas, the one most in use being a solution of whaleoil soap and quassia chips. The proportion varies from eight pounds of quassia chips and seven pounds soap to six pounds of quassia chips and 12 pounds of soap to the acre. I have met with good success with the last named proportion. The quassia chips should be fresh and finely cut, and the whaleoil soap must be strictly pure and of 80 per cent. test.

"Many growers have been disappointed in spraying because they used an inferior quality of material. Weigh out 20 pounds of chips and put them in a burlap sack, tie the end of the sack, sink it in a barrel of water and soak 24 hours. Then turn the water into a tank under which a fire can be built, put in the sack of chips and let them remain until the water commences to boil; then remove the sack of chips, from which all the strength has now been extracted. Now turn into the tank 40 pounds of soap, and boil until the soap is all dissolved; then add water until there are 50 gallons of the solution. In spraying, use five gallons of this solution to 35 gallons of water. It usually takes about

120 gallons of spray to go over an acre. With a roller sprayer and three men and two horses, eight acres can be sprayed a day, provided water is plentiful and near and the land reasonably level.

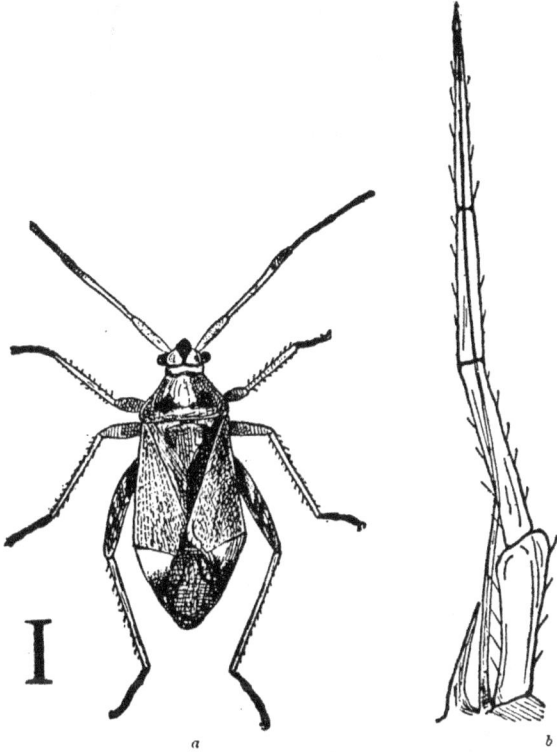

FIG. 73. NEEDLE-NOSED HOP BUG. (*Calocoris fulvomaculatus*).
b, Proboscis, greatly enlarged.

"Spraying should be commenced as soon as the lice make their appearance and should be continued as often as necessary, until the hops begin to burr, after which the spraying will do but little good, as the lice get inside the hop, where the spray cannot touch them. The number of times necessary to spray a hop yard depends upon the location, the density of the foliage of the vines and the weather during the months of June and July, as hop lice breed and increase very rapidly in damp, rainy weather and very slowly in hot, dry weather."

We find comparatively little in English or German methods to add to the foregoing, and the latter may well profit by American experience. Kentish planters take more care than others to prevent vines being too thick, and emphasize the fact that plenty of sunshine among the leaves is one of the best safeguards against lice and mold. A Washington farmer sets two or three rows of tobacco plants about his hop yards, which seem to attract the winged lice on their way from the plum tree to the vines, and cause them to die after feeding on the tobacco. In 1895, much injury was done in England by the needle-nosed hop bug (Fig. 73), which with its long, sharp proboscis (Fig. 73, *b*) punctured the tender parts of the vine (Fig. 74), not only drawing out the sap, but leaving a wound that bled and weakened the vine. The only thing that disabled them was 12 pounds of soft soap to 100 gallons of water (or of other washes) sprayed on a dull day, when the bugs are less active; they are best treated during the wingless stage early in the season.

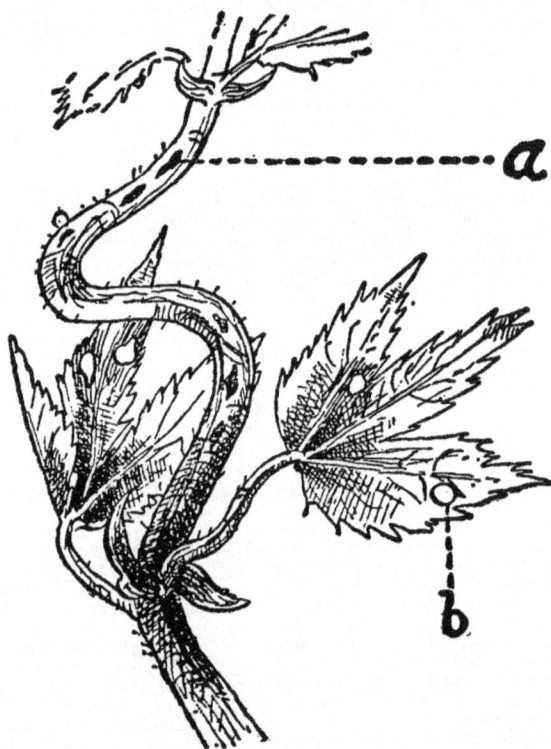

FIG. 74. HOP VINE AFFECTED BY NEEDLE-NOSED HOP BUG.
Showing "scars" on hop vine. *a*, Due to the *Calocoris*, *b*, holes in leaf due to punctured by the Anthocoris, a somewhat similar insect.

Prof. Jno. B. Smith lays as much stress now upon ladybirds for destroying hop lice as in 1887, when he first called attention to them and their larvae known as "niggers."

"Three species are found in abundance on hop vines. The most numerous is the two-spotted ladybird (*Adalia bipunctata*), a small red species, with two black spots on the wing covers. Next comes the nine-spotted ladybird (*Coccinella 9-notata*), a larger species, with nine black spots on its yellowish-red wing covers, and least numerous of all is the twice-stabbed ladybird (*Chilocorus bivulnerus*), smaller than either, entirely black, except two blood-red spots on the wing covers. The larvae of these species are all very much alike, and of an elongated, flat form, tapering toward the tip, with six legs; of a grayish-black color, spotted and marked with red or yellow. They are very active and very rapacious, feeding almost continually, and each larva destroys many lice before attaining maturity. When full grown, they attach themselves by the tail to a leaf, curl up into a round pellet, and in a few days transform into the perfect beetle, which also feeds on the aphis, but is not so voracious as the larva. There are several broods of the insect in the season, the last transforming into the perfect insect about the middle or toward the end of September.

"The beetles hibernate in crevices of fences, under bark of trees, or stones, or wherever else they can find shelter, and reappear in spring to continue the work where they left off the year before. Were it possible to preserve a sufficient number of these insects through the winter, so that a goodly number of them would be on hand in early spring, the lice would never become numerous enough to do injury; as it is, but few survive the winter, and before they become numerous the lice, propagating more rapidly, become so plentiful that they are beyond control. But, seriously, there is no reason why these coccinellids cannot be wintered. They become very numerous in fall, and several hundreds of them could be collected without difficulty, put into a large box with plenty of loose rubbish, and kept in some cool place not exposed to the fiercest cold, nor yet so warm as to cause them to become active—a barn or cellar would answer. The box should be covered so as to prevent the entrance of spiders which would feed on them. In spring, the box could be placed in the open air, and the insects would then scatter through the yards in search of suitable places to deposit eggs. I firmly believe that this could be done without much trouble, and that it would prove the best possible remedy to prevent the spread of or damage by the aphides."

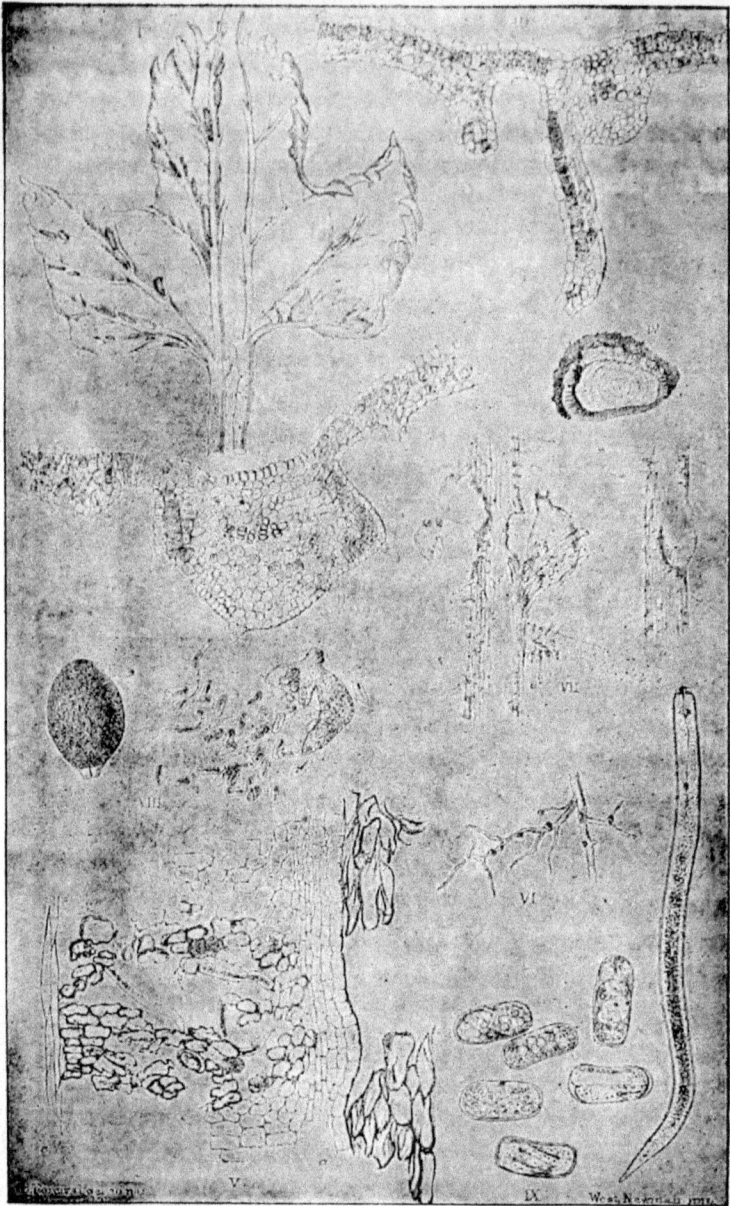

FIG. 75. AN EELWORM DISEASE OF HOPS.

The eelworm disease causes sickly looking bines and curling of the leaf. The trouble is due to a minute eelworm, which slits and injures the delicate rootlets. The accompanying cut (Fig. 75) is from *The Journal* of the Wye

Agricultural College for April, 1895. I shows the leaves, smaller in size than usual, *a* under, *b* upper, surface, showing characteristic curling of edges and puckering of veins; natural size. II, Cross-section of leaf, enlarged forty times, showing abnormal tissue. III, As in II, showing further growth of tissue at side of midrib. IV, Cross-section of root, natural size, and V, lengthwise section, both showing effect of stem-eelworm (*Tylenchus devastatrix*). VI, Hop rootlet with attached females of eelworm (*Heterodera schachtii*), almost natural size. VII, Magnified cross-section of rootlet, showing eelworms at work. VIII, *a*, female eelworm; *b*, ditto, broken, showing eggs and larvae. IX, Eggs at different stages, and the young worm, magnified 250 times. Dig out and burn infected plants. Lime, one-half ton per acre, or sulphate of potash, 200 to 400 pounds per acre, are the remedies suggested.

FUNGUS PESTS—BLIGHT, MOLDS, ETC.

"Fire blast" and "red rust" are not common in the United States, and the latter at least is due to an insect (the red spider) rather than to a fungus. Mildew and mold are also comparatively rare, though the attacks of lice often cause a blackened condition ignorantly called "mold." Mildew is one of the worst pests in England, and in damp seasons is almost equally destructive in Europe. The best account of the hop mold or its treatment is Percival's, in the Journal of the Wye (Kent) Agricultural College, under whose direction the test has been carefully studied and experimented with.

Symptoms—In the earliest stages, the mold is seen as small, light-colored patches, chiefly upon the upper surface of the leaves. If the nights are cold and damp and the hop plants in a backward or weakened condition, the patches soon increase in size, generally regularly from a center, so that the spots are approximately circular. As the patches increase to about one-eighth of an inch across, they become whiter in color (Fig. 76), and have a dusty or floury appearance. Fresh spots show themselves on the younger leaves and in bad cases the malady spreads from the lower leaves, where it is generally first seen, to those higher on the plant and even to the tender shoots and young hops. In all cases the plants suffer in health, but it is only when the tender shoots and young growth are attacked that serious damage is done. The young hops and tips of the laterals on the bine then lose their soft, succulent character and become deformed; the parts attacked dry up, and development is stopped. Often the white patches of mold do not spread; the spots lose their dusty appearance and vanish, leaving behind always a small yellow or brown dead place upon the leaf attacked. More frequently, however, if the mold is allowed to remain unchecked, and the weather is unfavorable to the growth of the hop plant, the patches, especially on the lower surface of the leaves and on the young hops, become covered with extremely small, dark, rusty-brown specks, and the white, dusty character of the spot gradually disappears.

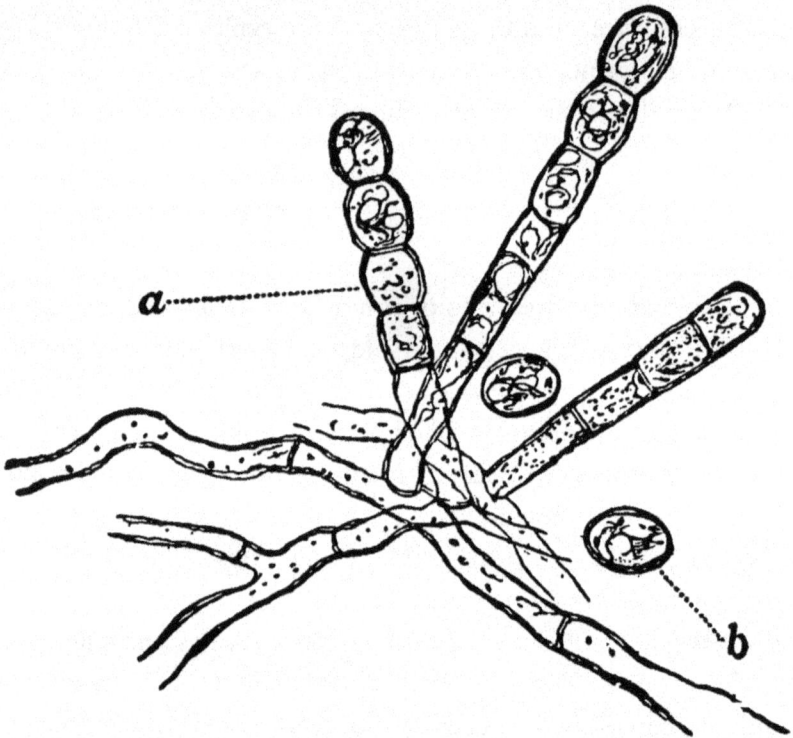

FIG. 76. THREADS OF HYPHAE OF HOP MOLD (Sphœrotheca castagnei). Summer, or active stage. *a*, Erect hypha, giving rise to chains of spores (*conidia*). *b*, Free "spore" (*conidium*).

The time at which mold is first observed varies with the season. Gardens once seriously attacked and neglected are always specially liable to an annual recurrence of the disease, unless measures are taken to get rid of the trouble.

Cause—The ordinary symptoms can readily be seen and followed by the naked eye, but the exact cause and process of development can only be appreciated fully after making observations with a good microscope. A mold spot in its early stages is then seen to be made up of a tangled mass of branching threads (hyphae). The threads are clear, transparent, hollow tubes, filled with living substance (protoplasm), and constitute the body or spawn (mycelium) of a fungus, known as *Sphacrotheca castagnei,* which is one of a large class known as "true," or "surface" mildews. Careful observations show that the threads are not merely resting on the surface of the leaf, like a tangled skein of cotton upon a table; they cannot be blown away or washed off, as at various points they are attached by short suckers (haustoria) which just penetrate into the substance of the leaf and serve the double purpose of holding the fungus in its place and acting as roots to suck up and convey the sap from the hop plant into its own body. The spawn of the fungus or mold thus lives upon the substances manufactured in the hop leaf, and is enabled to grow and spread. Not long after the fungus threads are established, they send up into the air short branches which give rise in a little time to rows or chains of minute oval-shaped bodies (Fig. 76*a*), known as spores (conidia). These spores, which for ordinary purposes may be looked upon as "seeds," are very small. They soon fall off the branch producing them (Fig. 76*b*), and by their number—many thousands in a single mold spot—increase the mealy appearance of the affected part. Being necessarily very light, many are blown about by the wind. Under proper conditions of moisture and temperature each one can germinate in a few hours and produce a small thread which fastens itself to the leaf of the hop and begins a new mold spot. We can thus understand how quickly and silently mold can spread in a garden. From one small patch several thousands can arise in a few days by the production and dissemination of these spores by the wind, much as thistles and other weeds may be spread about the country after seeds are produced. The germination of spores, and the growth of spawn producing more spores, can be repeated over and over again in a few days, and it is in this way that the mold spreads during the summer.

The spores and spawn, such as we have mentioned, are short lived and cannot exist through the winter. The fungus, however, in autumn, or when the leaf upon which it is living begins to die, produces upon its body of threads small round cases containing another kind of spore, which has the power of resting during winter, and when fully developed, these round cases (ascocarps, Fig. 77) are dark brown in color, or almost black, and give a rusty appearance to mold spots which have been allowed to develop unchecked, especially those on the underside of leaves and on the young hop cones. They are hollow, and constructed somewhat like a football, that is, one case inside another. The outer case is made of dark brown, strong material (Fig. 78*b*), and acts as a protective coat for the delicate, transparent, pear-shaped case (ascus) inside (Fig. 78*a*). The latter contains within it eight spores (Fig. 78*c*) (ascospores) about the shape and size as those produced upon the upright threads mentioned above, only they do not germinate so readily. These double cases, with their spores, are produced in large numbers in late summer or autumn in a badly affected garden, and fall upon the ground with the dead leaves, and the spores within are shot out into the air, and are carried to the young bines and leaves, which are then growing from the hill. Thus we see why it is that mold generally commences close to the ground and spreads upwards, and why there are "moldy places" in the gardens, where the malady begins almost every year.

FIG. 77. ASCOCARPS OF "HOP MOLD."
Autumn, or resting stage. Highly magnified.

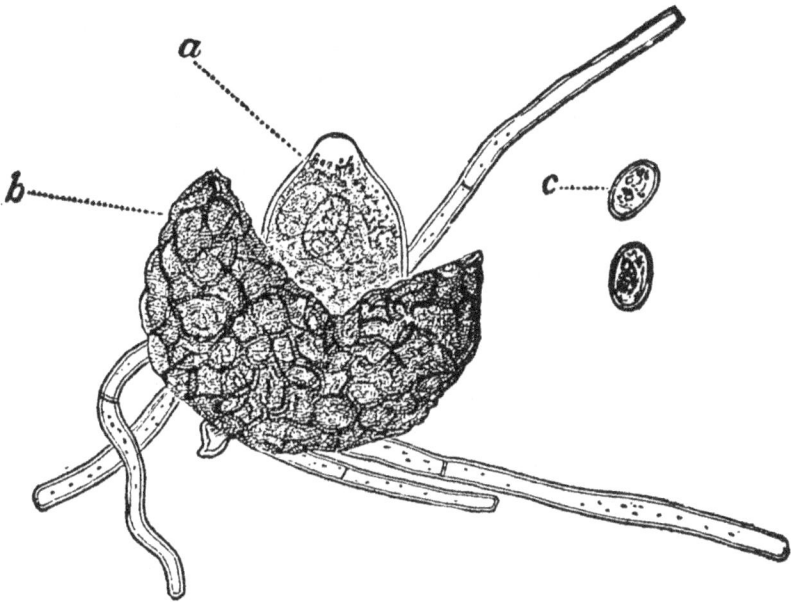

FIG. 78. SINGLE ASCOCARP.
b, Burst, showing ascus. *a*, With its contained ascospores. *c*, Free ascospores. Highly magnified.

Remedies—It will be understood from the foregoing account that we are dealing with a living pest, and that it is just as impossible to create "mold" as it is to manufacture aphides or lice. Various methods of cultivation and treatment of the hop plant and garden may be conducive to the growth and development of the "mold" fungus, but unless its spores are present or are introduced into the garden from outside, spots cannot occur. Whatever remedies are adopted, it is absolutely essential that they should be begun early, as neglect means failure. The pest soon becomes established and is then difficult to eradicate.

1. Although the complete destruction of hop mold is unattainable, every effort should be made to diminish its prevalence by burning all badly affected vines and leaves. This practice should especially be carried out in cases after a bad attack, where the hops have been not worth picking on account of moldiness. The vines should on no account be left lying about, as the spore cases are produced in thousands and fall upon the ground, only to remain a certain source of infection for succeeding years. The application of gypsum to the soil is said to be beneficial in such circumstances, but no trustworthy

135

experiments upon this matter have been carried out. Possibly lime might help to destroy the spore cases.

2. Certain varieties of hops seem to be specially liable to suffer from this trouble, but apart from possible inherent differences in the plants, more careful manuring should be adopted in order to produce a healthy growth. Excessive amounts of nitrogenous manures make the leaves more readily attackable by mold. Anything which reduces the vitality of the hop—such as cold and damp nights, long continued drouth, or wet weather and want of proper amount of sunshine and fresh air—indirectly aids mold in its ravages. It is generally in "housed-in" parts where the air is still and damp and where light does not easily penetrate that the worst effects are seen. Systems of training hops should aim at reducing these drawbacks to a minimum. Early training of the lower part of the bine diminishes the likelihood of attack from the soil and also allows of better air circulation.

3. The hop-mold fungus not only lives upon hops, but also upon many wild plants—groundsel, dandelion, strawberry, avens, meadow sweet, and many others. There is little doubt that it is from such sources outside the yard that many attacks of mold are begun.

4. The fungus lives and develops almost entirely upon the outside of the leaf, and on this account it would appear more easy to deal with it by means of washes and external applications of powdered substances than those cases like the potato disease, where the growth of the fungus goes on chiefly inside the leaf. The application by hand or bellows or by specially constructed sulphurators of finely powdered sulphur to the affected leaf is a remedy for mildews of various kinds, which has been employed for about half a century.

Mechanically powdered sulphur—roll brimstone reduced to a finely pulverized state, by hand or machinery—acts better than that form known as "flowers of sulphur," obtained by condensation of its vapor or by precipitation processes. In any case, the substance acts in two ways (1) as a fungicide—that is, a definite destroyer of the mold; and (2) as a protection against further attacks and spreading, as spores will not germinate upon a sulphured leaf.

It is chiefly as a protector that sulphur is so beneficial, and on this account every endeavor should be made to distribute it upon the youngest growth. As a direct fungicide, it possesses little effect, and even for this small benefit it must be repeated frequently where mold is bad. The best results with sulphur are observed when the temperature is above 78 degrees F., and it is, therefore, usually applied with success on clear, bright, hot days, usually in the middle of the day, or early morning when the leaves are partially damp with dew. In cold weather it is nearly useless, and in wet days the sulphur is soon washed off the leaf. The general explanation of its action is that the sulphur becomes oxidized, with the ultimate formation of sulphurous acid, and this latter substance is credited with the destroying effect upon the mold. Sulphurous acid, however, in exceedingly minute quantities, has a deleterious influence upon the hop leaf itself. Some experiments have indicated the formation of sulphureted hydrogen. The fact that sulphur works most beneficially on hot days, and also that the odor of a sulphured garden is not like that of either sulphur dioxide or sulphureted hydrogen, but resembles that of roll-brimstone itself, suggests that sulphur vapor may be the active agent. The possibility that the action is a mechanical one must also be borne in mind. Some authorities state almost any fine powder will do, that road scrapings, brick-dust, chalk, and ordinary flour work as well as sulphur.

5. Under the assumption that sulphur has some specific action upon the fungus, various soluble compounds containing the ingredient are employed, chiefly the sulphides of sodium, calcium, and potassium ("liver of sulphur"). These substances are, undoubtedly, of considerable use in checking and destroying molds of all kinds. They are readily soluble in water, and are generally applied in the ordinary washes of soft soap and quassia at the rate of 1½ or two pounds per 100 gallons of wash. A wash of this description, followed by an application of powdered sulphur, is perhaps the most effective and safe means known at present for an attack of mold. The alkaline sulphides in solution do not keep well, unless air is excluded from the vessels in which they are kept. Practically all mold washes have, as a basis, one or more of the above sulphides in conjunction with substances like glycerine, which tend to keep the wash upon the leaf till it has done its work, and which also prevent too rapid oxidation of the active ingredient. Although the preparation of washes is not difficult, a certain amount of chemical and botanical knowledge is essential to avoid damaging the plants, and until this is obtained it is perhaps the wisest plan to obtain chemicals or washes prepared ready for use from experienced manufacturers.

Many other substances, notably preparations of copper (Bordeaux mixture, "Fostite," talc and finely powdered copper sulphate), have a more certain effect in destroying mold, but the application to hops is scarcely feasible on account of their somewhat poisonous properties.

There are various definite chemical and physical differences between the cells and cell-walls of the hop leaf and the substance of "mold" fungus, and it should be possible to construct a wash or fungicide dependent upon these differences. This, however, remains to be accomplished, but until it is effected, washes and applications of powders will be uncertain in their action.

To the following may be added Whitehead's summary: The action of sulphur is materially influenced by conditions of weather. It is more powerful in heat, when volatilization takes place, and appears to be inoperative in dull, cold seasons. It should always be put on the plants in sunny, still weather, if possible, and as soon as they have got well over the poles. Another sulphuring should be given in about three weeks, and a third later on, especially if there are any traces of mildew upon the "burr." Should the fungus attack the cones when developing or when they are out, sulphur must be used again, though, if possible, this should be avoided, as brewers object sometimes to hops that have been sulphured when fully out. The quantity of sulphur applied for mildew varies from 40 to 60 pounds per acre, according to its lightness and quality. In France, very small quantities are put on at a time, with hand bellows, or soufflets. Sulphate of copper solutions have been tried by a few hop planters for mildew, but as yet no definite conclusions have been reached.

A hop blight more or less common in America, but not injurious is *Cylindrosporium humuli,* E & E. A prominent leaf spot on the hop is *Phyllosticta humuli,* S. & S. Halstead has found an anthracnose of the genus *Coletotrichum,* which causes blotches in the leaves, that turn white and fall away, leaving the foliage full of holes. In Oregon, the hop root bruised is apt to develop a fungus growth called "black knot." If cut off promptly it does no material damage, but if allowed to grow will cause the root to die in a couple of years.

Late frosts in spring may be partly guarded against by hilling up the young vines. Early frosts in fall may be mitigated by spraying with cold water, or by thick clouds of smoke from bonfires when frost threatens.

Hail, wind and flood often do much damage. If the vines are promptly trained up again, a surprising amount of the crop may be saved. There is a temptation to abandon a yard that is badly down, especially if the disaster occurs quite late in the season, but unless prices are very low, such a condition will often amply repay an effort to set it right.

FIG. 79. GROWING PINEAPPLES OR TOBACCO SHEDS IN FLORIDA.

Wind has a baneful effect upon hop plants when the burr is forming, and afterwards in all stages of the growth of the cones. It hinders their full development, and when they are getting ripe the heavy gales which often come towards the end of August make them brown by bruising them. In England, says Whitehead, many kinds of screens, or "lews," are adopted to lessen the force of the wind; some natural, as quick hedges, in parts of Kent, which grow as high as 20 to 25 feet in some districts, and rows of Lombardy and other kinds of poplar. Others are made of high poles set closely together, or of hop plants put as near to each other as possible, and trained up poles pitched close together around the outsides of the hop yard. Light cloth of a coarse mesh, made of cocoanut fiber, is stretched about twelve feet wide at about eight feet from the ground upon wires fixed to permanent poles, in those parts of the hop ground exposed to the prevailing wind. Screening in this way is expensive, but it is now adopted by most of the large Kentish planters.

Scorching is caused by the rays of a burning hot sun striking the plant when the air is perfectly calm or when there is not sufficient moisture in the ground to counterbalance evaporation through the leaves. The effects of scorching may be overcome by watering the plants morning and evening. The disease usually occurs when the hop is ripening. If after three or four days' watering it does not disappear, it is better to pick the cones at once rather than run the risk of losing the entire crop. It may be that scorching of hops in California or any hot climate might be avoided by partly shading the crop, according to the system in vogue in Florida for raising pineapples or cigar leaf tobacco of fancy quality, while the same method is being used in Florida to insure the orange crop against frost. These sheds (Fig. 79) might be modified so far as to be applicable to a solidly built trellis system, but it is doubtful if there is any practical virtue in the idea. Where shed culture of tobacco or pineapples is practiced, posts 3x3 inches are set in the soil 9x14 feet apart and 7 feet high. Stringers 1x8 are attached to the top of these posts the 14-foot way. These support the cover of the shed and should be braced at each post. The cover is made of 1x3 inch pine boards 18 feet long, laid flat and nailed to the stringers, leaving between these boards a 3-inch space.

FIG. 80. STEPLADDER.

CHAPTER XI.
Harvesting the Crop

THE time between spraying and harvesting is fully occupied in getting up wood for the dry kiln, hunting hop pickers and repairing hop kilns, boxes and hop sacks, getting supplies of hop cloth and sulphur, etc. Everything must be in readiness, so that the work of picking may go forward without interruption. Examine stoves and flues, and see that they are in order and clean. Have the pump or water supply near the hop kiln in good order. See that the bunk houses for pickers and sleeping quarters or tent rooms are clean, neat and orderly for their comfort; you lose nothing by this little attention to the comfort of your pickers. Provide a large trough with a stream of water going through it for washing and cleansing purposes for the help, and make suitable sanitary arrangements, not forgetting that children often have to go picking with their parents. Supply wagons, with meat, bread, groceries, etc., should visit the yards daily for the convenience of the pickers.

Get out the hop boxes, see that the handles are all right and that they are properly numbered, so that you can tell who picked the hops in any box. Have your hop tickets printed and ready for the pickers, and if you don't intend them to pass for credit outside the yard or to truck peddlers, have it so stated thereon. State the value of the ticket upon it, and have the same signed so that you may know your own tickets. Give all small details your attention before you commence picking, and arrange for your dryer, fireman, yard boss, pole puller, teamster and carrier. Have your wagon fixed up with hop rack—a frame level in surface, extending three feet over the reach and covered with boards, to carry about nine boxes of hops. Have your thermometers hung on a cord with pulley under the hop floor ceiling, and two lanterns with reflectors all clean and oiled ready for night work, placed outside, with a glass to show inside the ground floor. Sweep out your floors thoroughly, rake over the dirt floor of your stove room, and put your sulphur pan on a stand in the center of your stove room. Insure your kiln and hops in a good, reliable company for 30 days, and don't be caught by a fire, leaving you minus the crop and money.

FIG. 81. SCENE IN A KENTISH HOP YARD.

The date of picking depends upon the time of maturity. In Washington and Oregon, the harvest usually begins between September 1 and 10; in California, it is some days earlier; in New York, picking begins between August 22 and September 1; in England, about the same time, and in Europe, from August 15 to 25. Abroad, the harvest usually lasts through three weeks, as more care is taken there to have early, medium and late varieties to thus extend the harvest. In America, the bulk of the harvest is over in 12 to 15 days, but owing to very early or very backward yards, hop picking may cover over a month.

Hops are ripe and ready for picking when the seed becomes hard and the point of the cone closes up and the hop feels hard and solid when grasped in the hand and makes a rustling sound when touched. The hop is known to be ripe by the following signs: "During their development, the leaves turn from pale green to dark green, and assume a peculiar tint, indicating a less active circulation. The odor of the cones, previously herbaceous, becomes distinctly aromatic, and in some districts this odor is so strong as to inconvenience persons passing in the neighborhood of the hop yard. The umbels, or cones, change from pale green to a bright yellow and green color. They are closed and the green scales alternate with the yellow ones. If the scales are stripped off and rubbed between the hands, they impart a sticky sensation, but without any trace of moisture. The cones now possess considerable elasticity, as may be easily tested by the hands, and the extremities of the scales become brittle." Later the scales turn a deep red color, and the aromatic odor is still further accentuated, but the lupulin adheres less firmly and the quality of the scales deteriorates. The hop is then over-mature. But hops "go off" so fast that they often have to be picked before showing the signs of maturity, while if any light colored hops are desired, picking begins before they are ripe, "though this entails a loss of weight and brewing powers."

FIG. 82. INDIAN PICKERS' LODGES, CALIFORNIA.

It is therefore of utmost importance that the crop should be picked at the precise moment of maturity. If picked before they are perfectly ripe, the cones possess a beautiful green color, but lose more weight in drying than if picked at maturity; they keep badly and contain less lupulin, and the lupulin is less adherent. A large proportion of lupulin thus disappears during the different processes of picking and packing, and finally the grower has to submit to a reduction in price, corresponding to the loss in weight and quality, on hops picked too early. If picked too late, the cones lose their beautiful yellow-green color, so much esteemed by brewers.

In picking, the cones should not be detached in bunches, but two or three at most may be gathered together. The stalks, peduncles, may be cut with scissors, knives, or with finger nails, but care must be taken that the cut is clean. The stalk, or peduncle, left on the cone should be as short as possible to avoid an excess of bitterness in the beer, but in order to keep the cone intact and to prevent any loss, the stalk may be from a quarter to half an inch long. Not a single leaf should be mixed with the cones. Cones cut too long or mixed with leaves are inferior, and notwithstanding the slightly increased weight, the market value is considerably reduced. The increased weight due to the presence of long stalks and leaves cannot exceed 10 per cent. of the whole without rendering it unsalable, while the corresponding depreciation would exceed 25 per cent. on the price obtained if the crop had been properly picked. In this matter of picking clean, the utmost care must he insisted upon in America as it is in Germany and Kent. Picking into baskets holding three to seven bushels is preferred by some, to large boxes holding 20 bushels.

FIG. 83. HOP BIN FRAME, NEW YORK.

Indians, Japanese and Chinese, also whites, pick hops on the Pacific coast. In New York, pickers come largely from the villages and cities, as not enough help is available in the rural districts. The same is true in England, where 60,000 pickers come into the hop country from London. In Germany, the small yards and large families make pickers less of a problem. All ages and both sexes are everywhere employed, so that besides providing accommodations for them to be comfortable, due regard for the moralities of life should also be provided for. The absence of all restrictions, the indiscriminate huddling together of the sexes, the character of the work and the freedom that follows the day's tasks, all combine to tempt toward drink and cohabitation. Scandalous abuses have thus occurred in America, which make it increasingly difficult to get good help, and bring into the country at every harvest a body of people who care more for corrupt license than for the recreation or money afforded by hop picking. Sanitary and police authorities watch these matters closely in England. In New York, church missions work among the pickers commonly. If hop planters would co-operate with the best men and women among the pickers, an *esprit de corps* could be created that would insure against most of the possible evils referred to and aid in expediting the harvest.

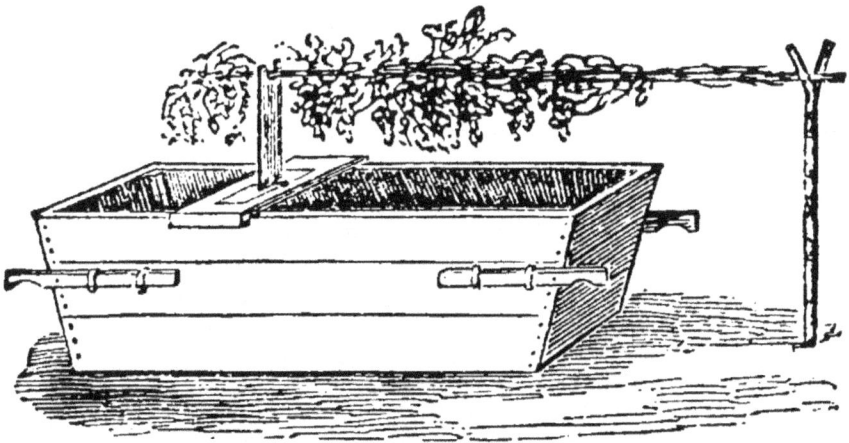

FIG. 84. PICKING BIN, NEW YORK.

RULES FOR PICKING AND PICKERS

In order that the harvest may be promptly and properly completed, it is necessary in handling large numbers of mixed help to have certain well-defined rules. These vary somewhat in different sections, but the most thorough and businesslike are those used by the Pleasanton Hop Company in handling its 300-acre yard in California, as follows:

NOTICE TO HOP PICKERS

YOU ARE ENGAGED UNDER THE CONDITIONS NAMED BELOW

1. We pay — per 100 lbs. to pickers who work throughout the harvest, from the time they begin to work.

2. All pickers must weigh in twice a day. The morning picking will be credited to the pickers' accounts, subject to these rules, the accounts to be adjusted at the end of harvest. Each afternoon's picking will be paid for with a negotiable check.

HOP PICKING CHECK

Good For Amount Indicated if Presented Before October 15 th.

NEGOTIABLE.

PLEASANTON HOP CO. CHOICE HOPS U.S.A. ALAMEDA COUNTY. CALIFORNIA.

№ 9006 G

This Check is issued for less than one dollar, and is not good if altered in any manner - Carbon Duplicate kept by the Company governs payment.

Date	Check No.	Section	Picker No	Weight	Amount
7/26	9006 G		84	100 Lbs	80 C's

PLEASANTON HOP CO.

FIG. 85.

146

3. Pickers quitting before the end of harvest will lose all their credits for mornings' pickings.

4. When hops are improperly picked, the picker will receive only one-half rate for the weighing in which such picking is found, and will be notified as promptly as possible.

5. When hops are found heated or discolored from long or heavy packing, the full amount due for picking such lots will be deducted from the picker's account. [This because the hops are thrown away.]

6. Pickers who are discharged for failing to do the work as required, will be paid only one-half of their credits.

7. Hops must be picked clean from vines and free from leaves and stems, and sacks must not contain any foreign matter.

8. Pickers must, with use of pole knives, themselves get down all portions of vines left clinging to the wires, and pick same before going to next vine. [The wire trellis system is used.]

9. Vines must not he pulled at the roots and must not be cut at the bottom.

10. Hops must not be packed tight in baskets or sacks. Pickers must not get on the hops in baskets or sacks or sit on same.

11. Hops must not be dragged on the ground on the vines or in sacks.

12. No picking to be done before daylight or after dusk. Picking hours to be the same for all and limited by the Pleasanton Hop Company as the curing process necessitates. No one allowed in the hop field outside of picking hours.

13. Pickers must bring their hops to scales in sacks weighing not over 80 lbs., get their weights and load them on wagons.

14. Each picker will be given a number which must always be shown, in plain view, for convenience of managing work in field. Numbers must be exhibited at scales, and to get sacks, and same must be surrendered at the office at final settlement.

15. All picking will be credited to the "number" presented with the hops at weighing. All empty sacks will be charged to the "number" presented when sacks are taken by pickers. The sacks brought to scales at weighing will be credited to the accompanying "number" and the picker will be charged 25 cents for each sack not returned.

16. Pickers getting baskets will be charged 25 cents for each one taken and credited with the same amount on returning basket before settlement.

17. The company does not furnish tents nor does it board the pickers.

18. No camping in hops.

19. No teams tied or fed on the hop held.

20. No smoking while picking or near others at work.

PICKERS, ATTENTION

As checks are cashed for exact amount indicated, pickers should see that trades people give them the full benefit thereof, either entire value of checks or make change to the exact cent.

WEIGHERS' INSTRUCTIONS

1. Weighers are also "field bosses," and as such foremen have full charge of their respective sections, they must see that all work is properly done and the picking rules adhered to.

2. To properly regulate the "give and take" of half pounds, will take the one-half pounds on morning's picking, and allow one-half pounds in afternoon's weighing. Weights must be accurate.

3. Non-negotiable, non-transferable memorandum weight credit slips (subject to the picking rules and all charges against the party to whom issued) will be given for morning's picking.

4. Negotiable checks are to be issued for the afternoon's picking, but no single check is to exceed 99 cents.

5. Where the afternoon weight of any picker calls for payment of more than 99 cents, then several checks are to be made out; where possible, for even amounts, making divisions by 100, 75, 50 or 25 lbs.

6. No checks are to be issued for less than 10 cents.

7. Weighers can hold checks for those pickers who do not wish to take them at the scales. These must be put in individual marked envelopes left at owner's risk in the office safe.

8. Issue no checks that show alterations of any kind. If mistakes are made in writing, mark "VOID" across detail line and turn in cancelled originals with duplicates.

9. Weighers will distribute sacks, charging same to pickers, and credit same when returned, noting against the scratched number, the letters "A" or "P," to indicate whether same were returned with morning or afternoon picking (and date of return when not brought in on the same day).

PICKERS WEIGHT MEMORANDUM
Retain This Ticket, it Must be Surrendered at Adjustmen,
Not Transferable.

This Slip has No Negotiable Value; it is simply intended as a weight credit tally, and is subject to the Company's Rules and all charges against the party to whom issued.

PLEASANTON HOP CO. CHOICE HOPS U.S.A. ALAMEDA COUNTY. CALIFORNIA

Nº 26 B

Date	Ticket No.	Section	Picker No	Pounds	Credit
7/26	26	B	42	75	62

PLEASANTON HOP CO.

FIG. 86.

10. Review "sack charges" daily, to make sure that pickers are not getting more sacks than they absolutely need.

11. Tag sacks with picker's number before weighing hops.

12. Report to office all charges for baskets and sacks not returned.

13. Report cause of quitting of such pickers who stop work or are discharged.

14. Weighers must see that all hops picked in the forenoon are "weighed in" by noon. Likewise, all afternoon pickings must be taken to scales when work is stopped. No hops to remain in baskets or sacks during the lunch hour or over night.

15. Arrange "carbon duplicates" according to picker's number, and file each morning's and afternoon's tickets in distinct bunches for reference.

The above rules at first reading may seem severe, but a thorough system of direction in the field and an occasional fine reported from the kilns when the hops are dumped against the "number" of a carelessly picked sack are a salutary lesson to an entire section and therefore few fines are necessary. Rule 3 may seem arduous, but as employment lists are closed when the company has a proper complement of pickers, it must insist that those who engage remain

149

until the entire crop is harvested. Few wish to quit and the rule is of course not enforced where there is a good cause for quitting, in which event, the picker is paid in full. So, too, Rule 6 is dependent upon the circumstances of discharge.

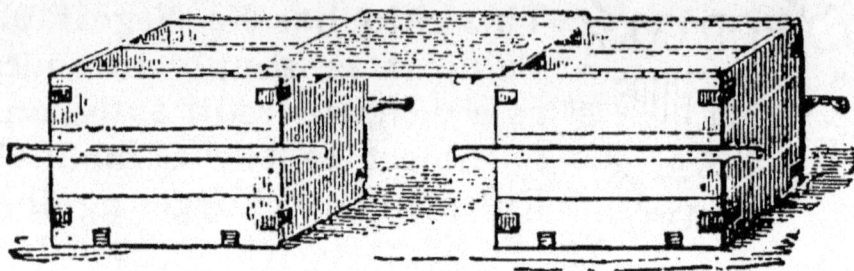

FIG. 87. "SET" FOR FOUR PICKERS.

The price for picking agreed upon by the growers is generally based upon the prospective value of the crop. But these agreements do not always hold good, as there is apt to be a strife to get pickers after harvest commences, as but few growers get all the pickers they have engaged and there is always some one short of pickers, and for the sake of getting their crop harvested quickly they will offer an advance above their neighbor. The other growers will be compelled to meet this advance or lose a part of their pickers. To such an extent has this been carried on that in Washington during 1896-7 many growers paid as much and in many cases more for picking than the crop brought them when sold. This trouble prevails more or less everywhere. Many hop yards are managed by renters, who harvest their crops upon money borrowed from the banks upon the owner's indorsement, and therefore if hops are worth only a small margin above the cost of harvesting, renters, seeing they may be unable to make anything, and having nothing to lose, do not care how much the harvesting costs.

In America, from 70 cents to $1.25 per 100 pounds of green hops has been the range of late years, mostly 80 cents to $1, but $1.25 may be paid when hop values are up and pickers scarce. For the '97 crop, the Pleasanton Company paid 80 cents the first week, 90 cents the second, and $1 the third week, against 70, 80 and 90 cents the previous season, whereas $1 straight may be paid in a prosperous season. In England, pickers are paid 2½ cents to 6 cents per bushel, averaging 4 cents; as a bushel weighs about five pounds, these prices are equivalent to from 50 cents to $1.20 per 100 pounds, or an average of 80 cents. In Germany, cost of picking is still less, and in many cases quite nominal.

FIG. 88. WEIGHING HOPS (California).

An average picker will pick from 80 to 125 pounds of hops per day—6 a.m. to 6 p.m.,—fast ones picking as high as 200 pounds under favorable conditions, but rapidity is apt to be at the expense of cleanliness. A 100-pound box of green hops will shrink to about 25 pounds of cured hops. Careful data from Pleasanton result in this statement: "As the hops grow riper, pickers cannot get such good results, whereas the more mature hops lose less weight on the kilns and therefore better pay is possible. In other words, the hops grow lighter in weight on the vines and dry out less on the kilns as the season advances, and while it requires about 3¾ pounds of green hops at the earlier stages of picking, hardly 3¼ pounds are necessary toward the close of harvest to make one pound of dried hops, or an average of about 3½ pounds, when the crop is properly handled. This at the normal price of $1 per 100 pounds for green hops would make the picking alone of one pound of dried hops represent 3½ cents."

In handling a large harvest, as at Pleasanton, the help are divided off into gangs or sections of 200 pickers each (in '97 eight such gangs were needed, "A" to "H" inclusive), which are in charge of the "weigher" and an assistant known as the "field boss." The weigher, as his title implies, weighs the hops, which are brought to the scales by pickers, and issues checks (Figs. 85 and 86). He also distributes baskets and sacks and makes all reports to the office. He is the real "field boss." His assistant, the "acting field boss," circulates among the pickers to direct their work and see that the rules are strictly observed. On a smaller scale, the same general practice is followed elsewhere.

When the picker's box is full it is delivered to the weighman, or measurer, who gives the picker a check for it and retains the duplicate stub for the bookkeeper, who compares same with the record of receipts at the kiln. In small yards, tab is kept in a book by the measurer. The best system to avoid all possibility of error is to have a paying machine like a cash register. The tickets are issued from automatic triplicating machines, the printed form (Fig. 86) going to pickers, the duplicate being retained by weigher for reference, while a secret triplicate roll remains locked in the machine, which can be opened only in the office and from which postings are made to individual accounts and from which also daily recapitulations are made on adding machines.

By the high trellis system (Fig. 48), the pickers cut the strings and vines off from the wires 18 feet above ground by means of a knife on a long pole; then pick off the hops from the reclining vines, which can be readily handled. Sometimes in England and Germany, the hop vine is taken down and stretched on hooks in the posts only five feet above ground. On the short pole system, cut the vine just below the hops and in the string above, slide the bearing vine down the poles, then pick. On the long-pole system, the pole-puller will cut the vine two to five feet above ground, and draw the pole gently, laying it on a crotch (Fig. 83) for the picker—not over the box, as the leaves would drop in fast. The simplest means of taking out a pole is to pass a chain or rope around it close to the ground, through which a lever is passed, and with a block of wood as a fulcrum the lever is thrust deeper as the pole is raised.

Numerous efforts have been made to perfect a hop-picking machine. It is only a question of time before some device of the kind will become practical, if, indeed, one or two machines are not already worthy of general introduction. They will doubtless be so altered and improved, however, that it hardly seems expedient to devote more space to them here.

FIG. 89. ELEVATING HOPS TO KILN.

When the harvest is completed, the poles should be carefully piled or stacked, all vines and strings collected and burned to destroy eggs of insects or fungi (or the vines may be used as stated on Pages 21 and 68), and the plants dunged with stable manure if the land is at all poor. Many are careful not to cut the vines at the bottom, where the trellis system is used, but let them remain until killed by frost, in order to mature the root, when the vines are cut and gathered. Sometimes the vines are cut into short pieces and plowed under.

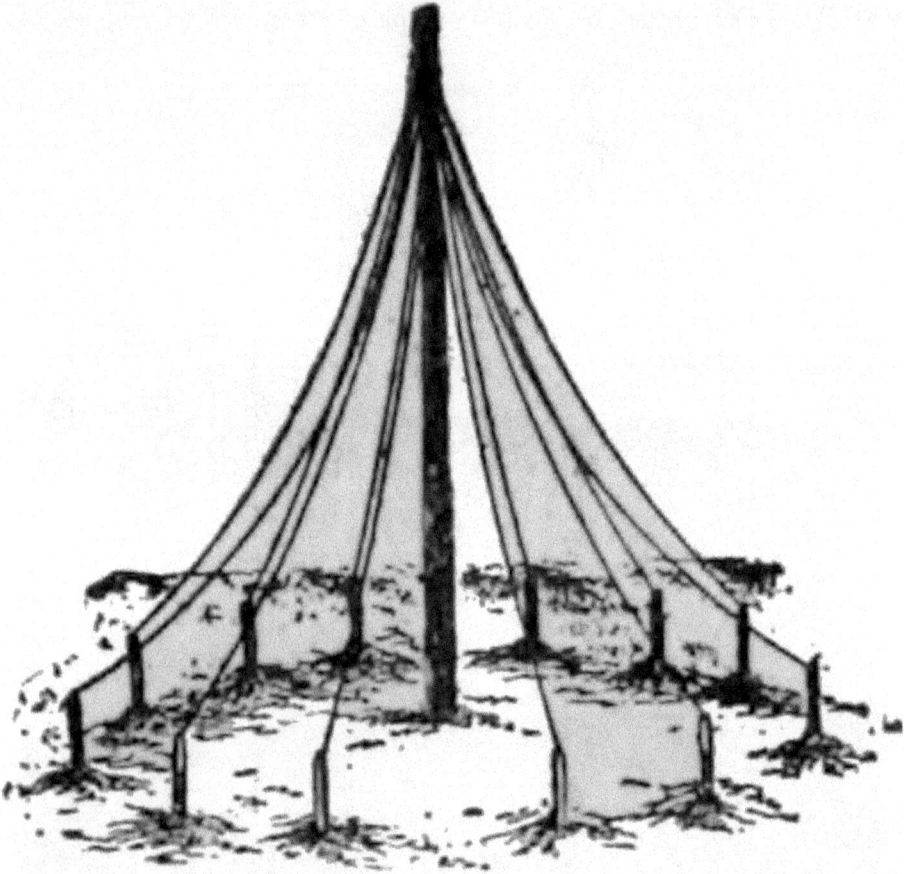

FIG. 90. TENT TRAINING.

FIG. 91. IMPROVED ENGLISH OAST.
This represents the most modern construction and all the latest improvements in vogue in England. Erected for Mr. W. Lillywhite, Wincheap Farm, Canterbury, Kent. From a photograph taken for this book by R. M. Elvy.

CHAPTER XII.
KILNS FOR CURING HOPS

IN Germany, the growers merely air-dry or sun-dry their hops. This is partly because the average grower has too small a hop yard to warrant a kiln, and also because the German trade prefers the present system. In Germany, if a specially fitted drying room has been set apart for the purpose, the large baskets or sacks are at once carried there and emptied, but if no such room is available, the hops are deposited upon screens exposed to the sun but sheltered from the wind, and in the evening, they are removed to an airy barn and at once spread on the floor. When hops are dried by this latter method, the walls and roof should be thoroughly cleaned and dusted beforehand, and the floor well scrubbed with soap and water, so that all dirt, vermin and bad smells are removed. Drying is done by aeration, and dust must be carefully excluded. It cannot be urged too forcibly that lofts or barns in which hops are dried must be perfectly clean and sweet. In any case, the large baskets, filled or not, must be emptied twice during the day, for if the hops are left closely packed together for more than six hours, fermentation sets in and the quality deteriorates.

It is claimed that this "natural cure" preserves far more of the essential oils and other brewing principles than is possible by the artificial hot-air cure in England and America, and that this accounts in part for the peculiarities of Spalt hops that command such extraordinary prices. The dealers buy the hops loose from the grower, sack them, carefully assort the hops, putting all of one color and strength together before bleaching them with sulphur; single firms thus handle and bleach 20,000 bales or more. Spalt hops are never bleached.

In England and America, curing is done in specially constructed houses, in which temperature, moisture and sulphur fumes can be regulated to a nicety. The construction of these curing houses will be first described.

are well and briefly described by Whitehead:

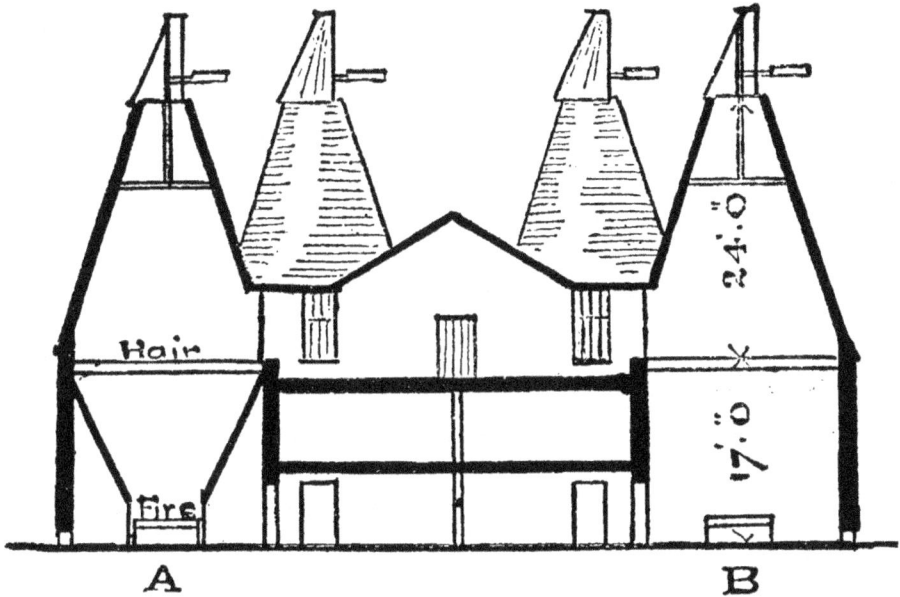

FIG. 92. SECTION OF GROUP OF KILNS AND COOLING ROOM.

"The kilns for drying hops are of simple construction, being occasionally square, but more frequently round, chambers, from 16 to 20 feet in diameter, with stoves or fireplaces in them, and from 14 to 18 feet high; at this height a floor of narrow joists, or oast laths, an inch and a-half or so apart, is laid over the chamber. At this point the sharply sloping roof commences, being carried up to an apex with a circular aperture of from two to three feet, upon which a cowl is fixed. The roof is from 20 to 26 feet high. A section of a kiln is given in Fig. 92, B, in which the relative height of the various parts is indicated. The kiln, or chamber, is in some cases merely a room with open iron stoves in it, as shown in the two lower kilns of the ground plan D in Fig. 93 and in Fig. 92, B, having holes at intervals in the walls, just above the ground level to allow the admission of cold drafts to drive up the hot air through the hops above. Over the open stoves, iron plates are hung, five or six feet from the floor, to break and distribute the volume of heat from the stoves. The cold air currents can be regulated by shutters over the draft holes. It is better that the stoves in the chambers should be set in brickwork, forming an inner circle (Fig. 92, A, and the two upper kilns in Fig. 93), so that the hot air is more concentrated, while the cold drafts do not mingle with it directly and diminish its heat. Upon the floor of joists or oast laths horsehair cloth is nailed to prevent the hop dust from falling through, and to keep the hops from burning (Fig. 92, A)."

The author begs to remark that such "oast houses" are regarded by progressive American hop growers as fifty years behind the times and afford few, if any, tests of value, except of how not to do it.

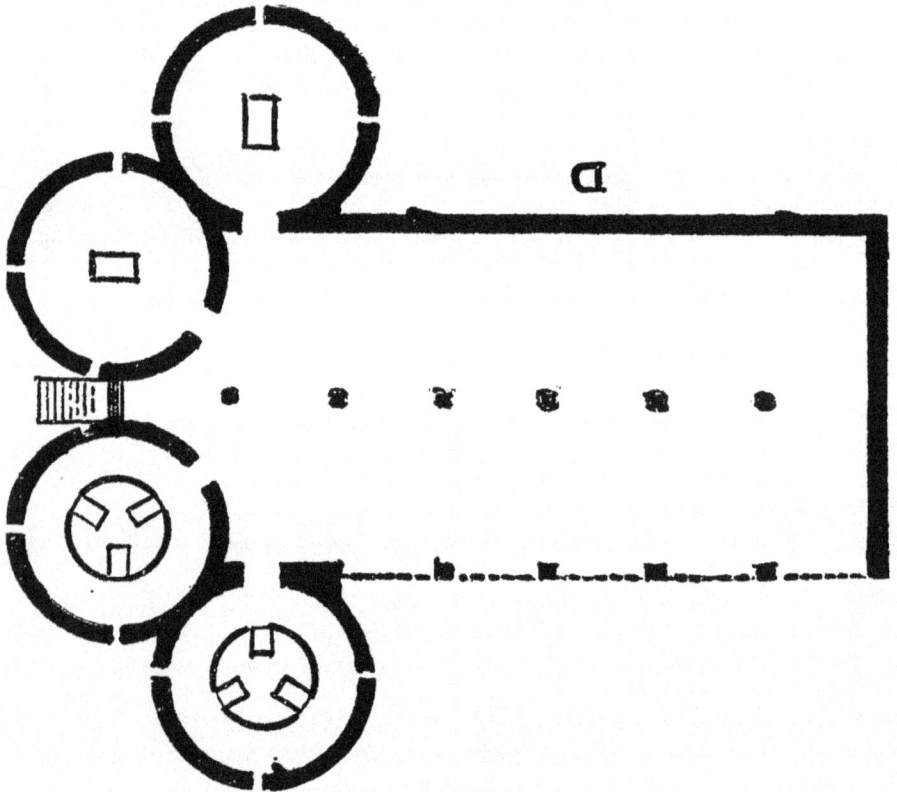

FIG. 93. GROUND FLOOR OF KILNS AND COOLING ROOM.

are of various kinds. Some are very old, but those recently built embody many of the improvements seen in the new curing houses on the Pacific coast.

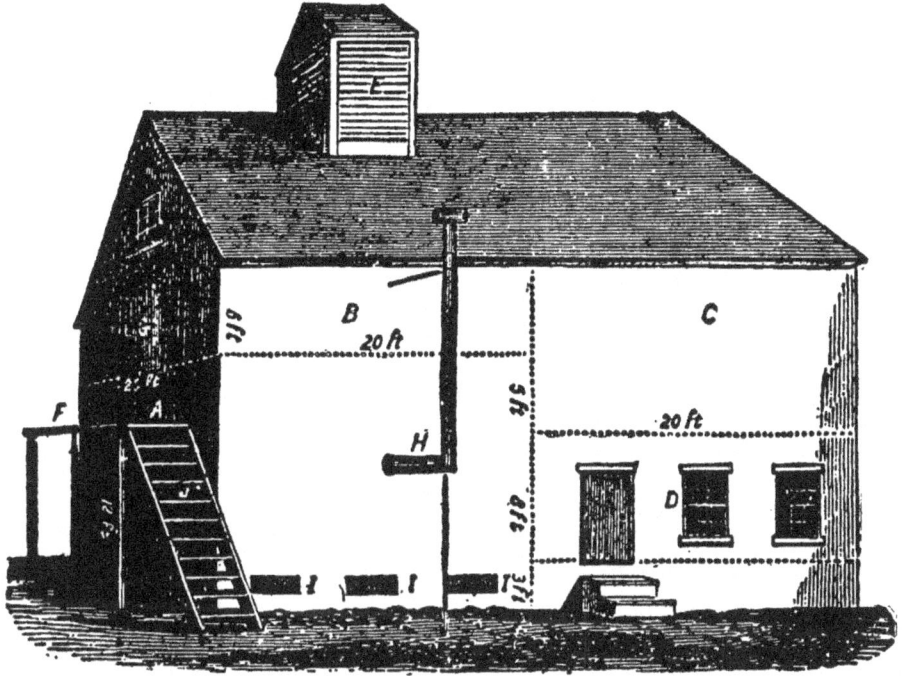

FIG. 94. ELEVATION OF THE COMMON HOP KILN.

A, Stove room, with stone, brick or plastered walls, but no floor; *B,* drying room; *C,* store room, which has a window in the end, not shown, with tight shutters; *E,* ventilator; *F,* platform from which to pass up the bags of green hops; *G,* door into drying room; *H,* pipe, or smoke stack from stove, which is to be taken down when not in use; *I,* air holes; *J* stairs to platform. The usual dimensions are marked on the diagram, but these may be altered to suit the size of the yard.

A familiar New York style is shown in Figs. 94, 95, 96, 97 and 98. The house is usually divided into four rooms. The stove room, where fire is made, should be not less than 14 feet high, and 16 or 18 feet is better, with stone or brick walls, and no floor. If the walls are of wood, they must be plastered to the top of the room. At the bottom of the walls there should be six air holes, one by three feet, with doors to close them tight when necessary; and if the kiln is very large, there must be more than six. The stoves, usually two, are large enough to take in three-foot wood, with grate bars at the bottom, and very large doors; the pipes are carried pipe is often run several feet from the building, and turned up like the once or twice across the room, as near the level of the top of the stove as possible, and then go into a chimney on the outside of the building. The smokestack of a steam boiler, to make a good draft.

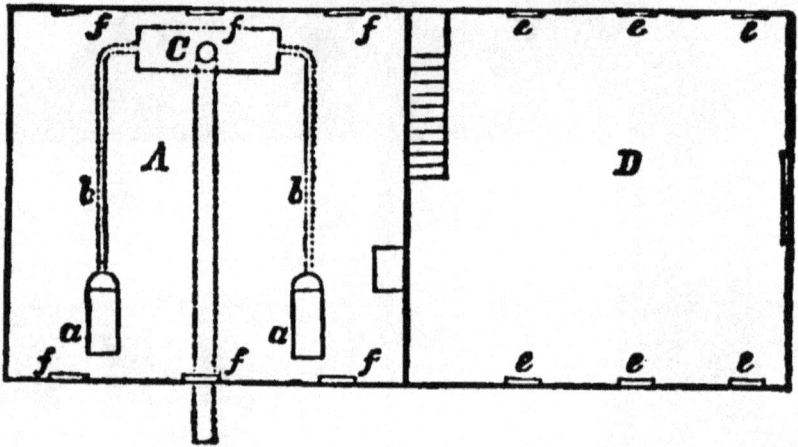

FIG. 95. GROUND PLAN OF HOP KILN.

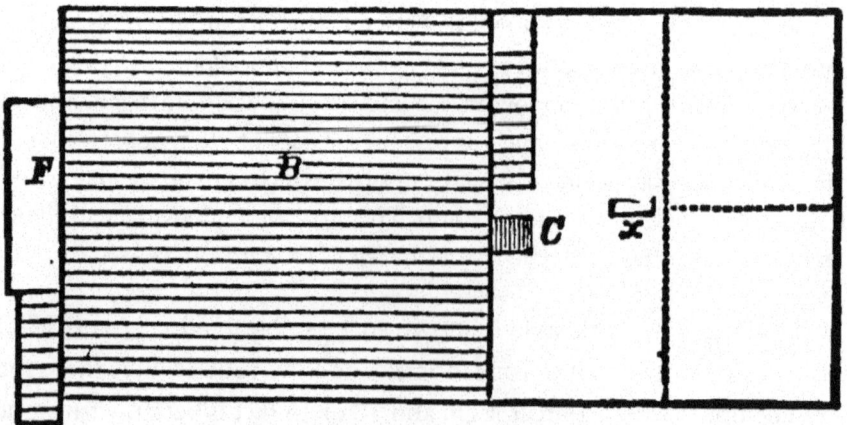

FIG. 96. SECOND FLOOR OF HOP KILN.

160

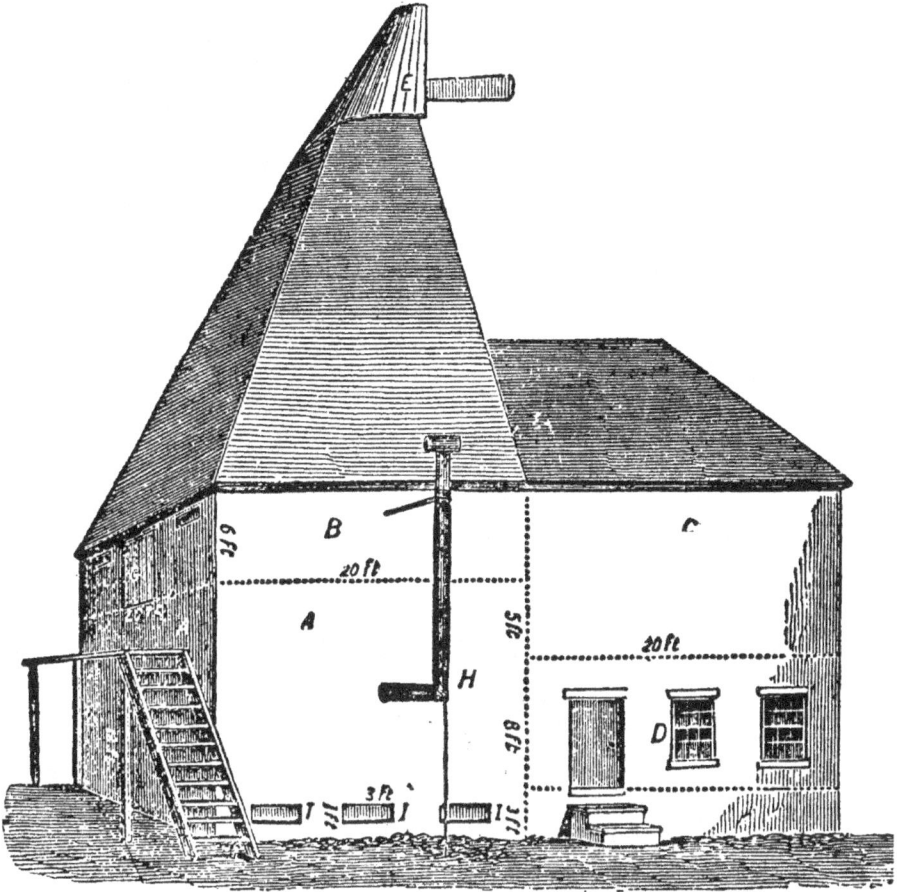

FIG. 97. DRAFT HOP KILN.
The figures give the dimensions, and the letters indicate the same parts as in Fig. 94.

There is a door from the stove room into the baling room, with a light of glass, so that the man who attends the drying may see the state of the fires without going in; a thermometer on the inside shows the degree of heat at a glance. The drying room is over the stove room. Usually there are joists laid across the top of the stove room, and wooden slats, one inch by two, are laid on them on edge, two and a half inches apart. On this there is laid a carpet, usually made of flax or hemp, with small threads, twisted hard and woven loosely, so that the spaces between them are about one-sixteenth of an inch or more, allowing air to pass through freely. It should never be of cotton. The roof should be carried up very high, so as to have the ventilator as high as possible, and make a better draft to the kiln. This is made with a cowl, which turns by the wind, or a slate ventilator is used, arranged so as to keep out the rain, while the air can pass up freely. The store room is next the drying room, but the floor is from three to eight feet lower, so as to make plenty of room to store hops in bulk until they are ready to press. It should have but one window, which should have a shutter to keep the room dark while the hops are in it. They will turn brown if exposed to light. Have boards to set up, and make the end of the store room farthest from the drying room into one or two large bins, so that any damaged hops can be kept separate. Under the store room is the baling room; it has a tight floor, and is used to bale the hops, store the hop press, together with any tools not in use in the yard.

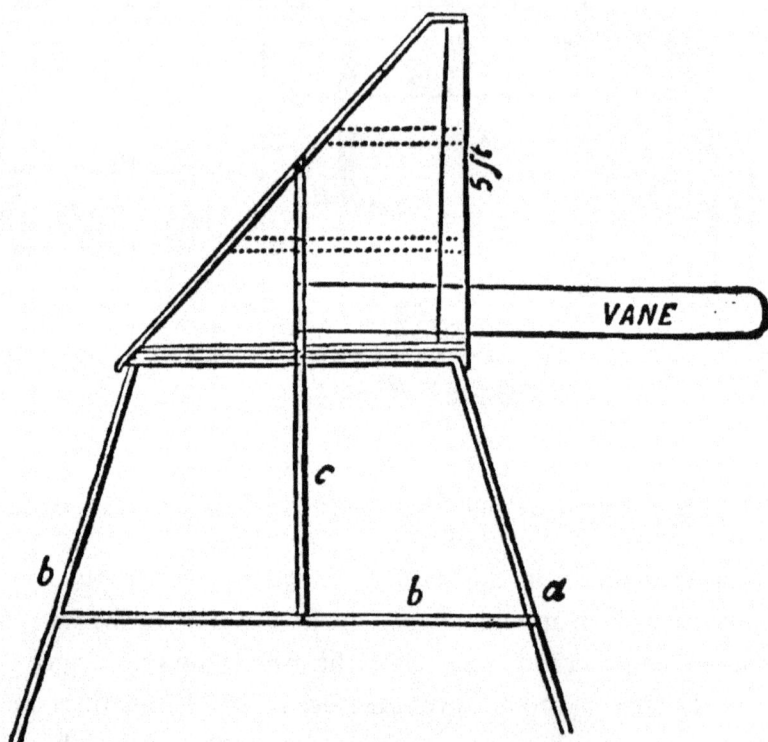

FIG. 98. SECTION OF COWL TO DRAFT-KILN.

a, Continuation of roof; *b*, 3x5 joist framed into rafters of roof; *c*, post, 3x3, framed into cowl, and movable upon an iron pin at bottom, which rests on *b*. The cowl shuts over the termination of the roof, and projects over it about two inches.

Another and more modern plan is illustrated in Figs. 106, 107 and 108. The size given is large enough for a yard of four or five acres. It should be set in a side-hill, if possible, otherwise much hard labor would always be required to get the hops up to the kiln. The hop house here described is 22x32 feet, with a kiln 16x16 feet, and a walk entirely around it. The stove room is 12x22 and two and one-half feet lower than the level of the kiln, which is 11 feet from the ground. The joists (j, j) over the stove room are two by seven inches, upon which rest the slats (s,s), one and one-half inches square and four inches apart. These support the strong linen strainer cloth, which is fastened to the side boards of the kiln, by small hooks. At the openings, where the hops are shoveled off, the cloth should be nailed down with small tacks. In Fig. 127 one corner of the kiln is shown, partly covered by joists, slats, and cloth. The dry room should be double-boarded or lathed and plastered all around to the eaves, and next the store room to the ridge. There should be a ventilator directly over the kiln. The store room should be boarded on the inside, next the dry room, and a space left for cool air to pass up, as indicated by the arrows in Fig. 108. This prevents the hops in the store room being dried continually by contact with the dry room. A hole (H) is left in the floor of the store room, in which a bottomless bag is fixed to conduct the hops into the box in pressing.

The stove room should be double-boarded outside, and double-boarded or lathed and plastered inside, and supplied with convenient air holes at the bottom on all sides, which may be opened or be closed up at pleasure. The stove is made expressly for drying hops. The bottom is simply a grate, so that the draft is directly under the fire, and consequently greater. The pipe (p), which should be seven inches in diameter, rises from the stove to the height of five or six feet from the ground, then passes horizontally into a drum, 12 or 14 inches in diameter and three feet long, thence as indicated by the arrows in Fig. 107, rising gradually, as seen in Fig. 108, until it reaches the chimney about four feet from the cloth. Such an arrangement of pipe keeps all the heat where it is needed, and, of course, saves fuel.

The press room should be at least seven feet from the floor to the beam in which the screws are set. The beam, ten by twelve inches, may also serve as a support for the floor of the store room. The bed-sill is of similar dimensions, and connected with the beam by two half-inch iron rods, seen in Fig. 108. In Fig. 107, B, is seen the bottom plank of the box, which is seventeen and three-quarter inches wide and six feet long, and is pinned to the sills. The side planks (c, Fig. 108) are of the same length as the bottom, and two feet wide, grooved near the ends to receive the end pieces. The length of the box inside is five feet. The top plank (d), one foot wide, is held in place by the ends of the tenons on the posts g. The cloth used for baling hops is about forty-four inches wide, and five yards is sufficient for a bale.

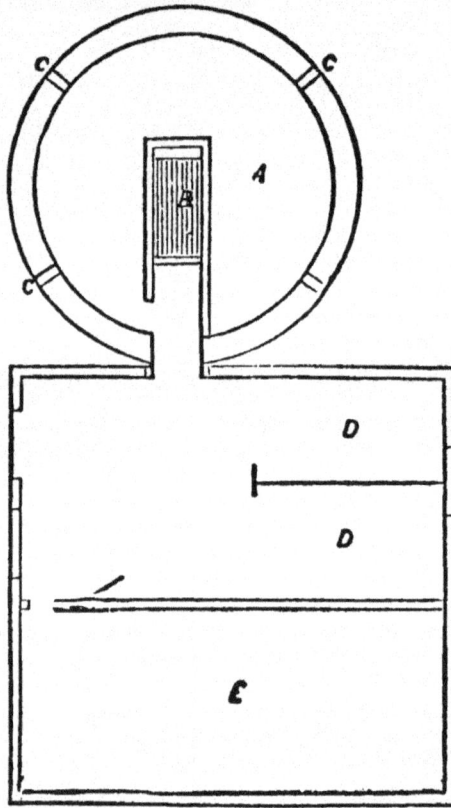

FIG. 99. GROUND PLAN OF KILN.
A, stove room; B, stove; C, C, draft holes; D, D, coal bins; E, press room.

The circular oast house is also employed, like the photograph (Fig. 123), and the floor plans in Fig. 99 and 100. A circular or square wall of brick, one foot thick, about 20 in diameter, is carried up to the height of 12 feet; then joists are placed in the wall at the height of 11 feet, across which are placed strips two inches square, and nine inches apart. Over these is spread a strong cloth made of horse hair. Figure 100 shows a plan of the drying floor, capacity

164

35 to 50 bushels. The wall is carried about two feet higher, and plates are placed upon it, and terminated by a sharp wooden roof. At the top of the roof should be a hole about five feet in diameter, around which is placed a circular plate somewhat larger on the outside than the hole itself. Upon this plate is placed a cowl, to keep out the rain and let off the vapor. It turns with the wind. On the ground floor is the furnace. A door connects the kiln with the storage room below and the chambers above, for receiving, cooling and packing the hops. The furnace is built so that the heat rises from the center. A wall two feet high is raised, upon which is placed an iron grate, three feet wide and four feet long. The wall is carried a few bricks higher, solid, after which it is raised in open work two feet higher, the bricks lapping over each other about two inches. The two sides and back end being built, the top is covered by flat tile, supported by iron bars, laid across. A ground plan is given in Fig. 99. A double kiln of this nature is shown in Figs. 110 and 111.

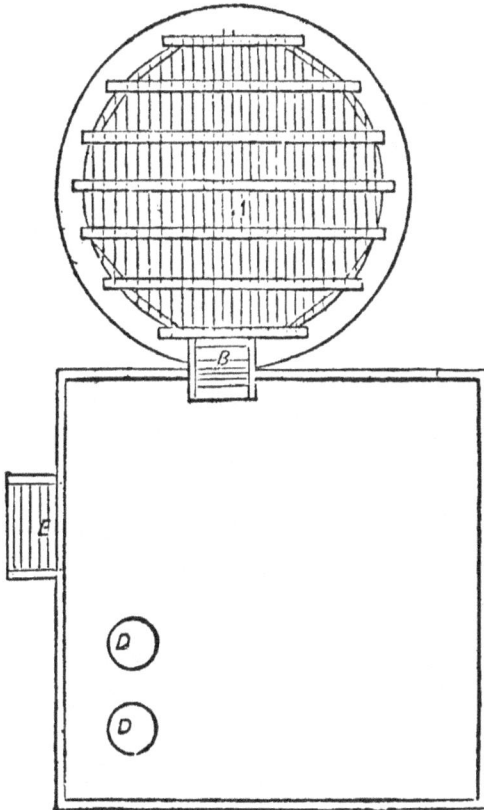

FIG. 100. PLAN OF DRYING FLOOR.

165

HOP KILNS ON THE PACIFIC COAST

Many of these have been built since 1890, and are designed to do their work with the utmost perfection and economy of capital, labor, fuel, and maintenance. These objects have been sought with special care by the Pleasanton Hop Company, whose buildings embody many features suggested as desirable by science and practice in all parts of the hop-growing world. This model hop-curing establishment is described in detail in the sketches, plans and photographs, Figs. 115 to 122, inclusive. This establishment now has 12 kilns, each 30x30 feet, all connected by over-head trestles with the six bins in each of the two large cooling rooms or warehouses. The cars in which the hops are carried from the kilns to the cooling bins are 30 feet long by 12 feet wide, big enough to take an entire "flooring" at once. The cars have movable sides and Λ-shaped bottoms, so that the hops can slide from car to floor in any direction wanted without being rehandled, which also saves breaking. Indeed, handling is avoided throughout the whole process, so as to secure the whole-berried or beaky hops desired by brewers. The hops remain undisturbed in the cooling rooms until ready for baling, and require about a week to cool off. The large power press in each cooler is so constructed that trampling the hops is unnecessary.

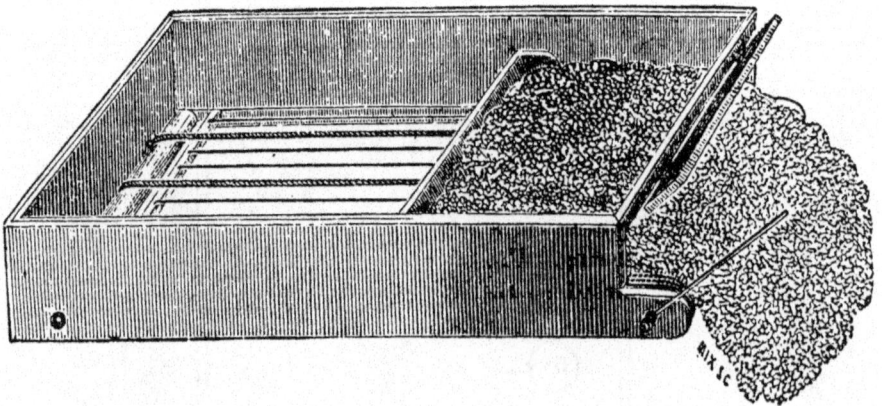

FIG. 101. IMPROVED FRANCE KILN.

The principles of construction outlined are also applied in Oregon, where kilns are usually 24x24 feet; also in Washington, where a few are 22x26 feet, rarely 30x30 feet, and a few old kilns are 16 feet square for 10 acres. The foundation sills (*b*, Fig. 102), of 6x6 inch stuff, rest on six by six pieces (*c*), two and one-half to three feet above a stout sill (*d*) on the ground, with the space below the sills open to give plenty of draft to the building. The studding (*a*) is of two by six, sheathed outside with rustic boards, inside lathed and plastered to the eaves. The roof is a half-pitch hip-roof, the rafters ceiled up with matched boards to the ventilator, which should be five feet square on the inside, and 12 or 14 feet high, and boxed up to within three feet of the top, with swinging doors, to close at pleasure.

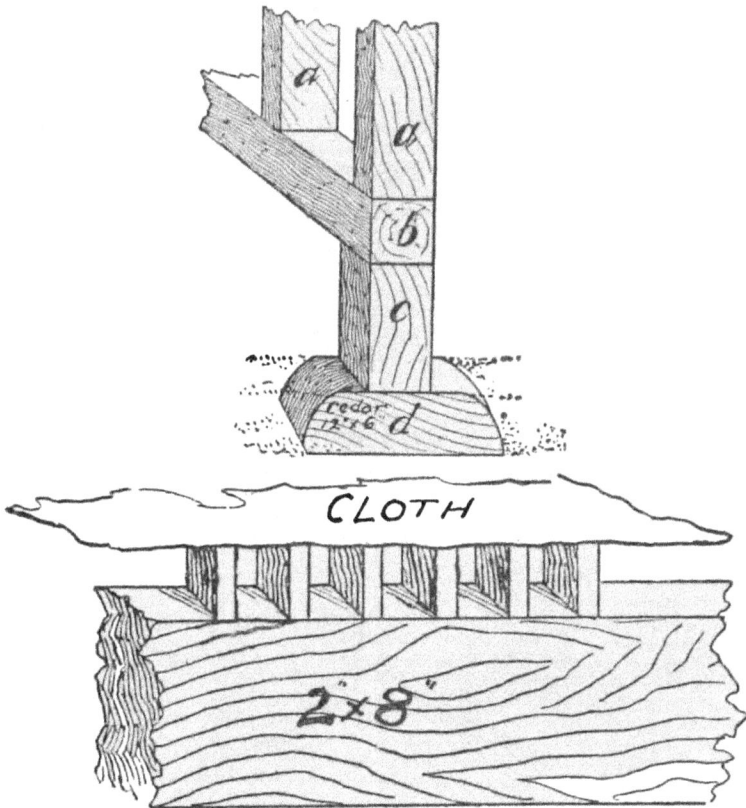

FIG. 102. DETAILS OF KILN CONSTRUCTION.

The hop kiln floor is usually 16 feet above the earth, or four feet below the plates, as too large an air space above the hops tends to check the draft so necessary to carry off the moist vapor and steam. The floor joists are two by

167

eight, resting upon a two by eight plate let into each stud one inch, and well spiked. Rough boards are nailed down and covered with one by four inch boards to make the floor. Hop-floor laths an inch thick and two inches wide, sized and with one edge rounded, are placed on the floor about an inch apart (Fig. 102), upon which in due time the hop cloth or carpet is stretched. In the France kiln the cloth is stretched on wires, and is rolled off by a shaft in the store room, so that all the hops are taken off in five minutes and the carpet put back ready for a new change without losing the heat or letting the fire go down. An improvement on this device is shown in Fig. 101. The hops are put on from a movable walk—a plank two and one-half feet above the carpet, supported from the rafters by wire suspension rods—and when the hops are on, the plank is turned on edge.

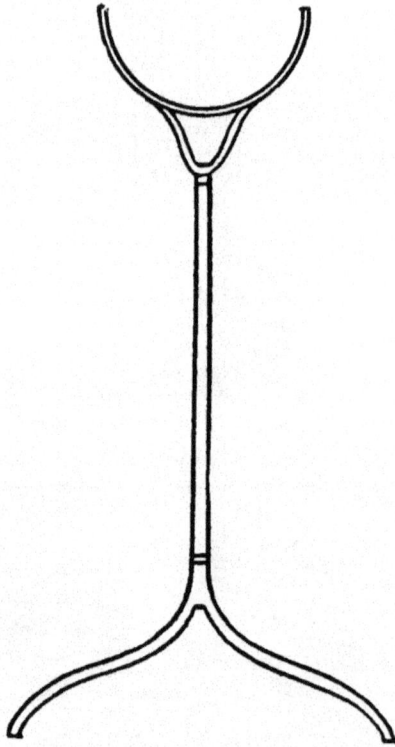

FIG. 103. SUPPORT FOR HOT AIR PIPE.

The iron drying stove (Fig. 104), big enough to take in four-foot wood, is set in brickwork, to prevent fire, the underside of the stove not lower than the sill. A 12-inch iron pipe runs up from the stove, breaking into a T (*a a a*, Fig.

104), the two arms gradually rising on supports (Fig. 103), but being kept about three feet from the walls, to avoid fire; when the pipes reach the other end of the room, they are joined by a T and carried into the chimney, built outside, which has a 12x12-inch flue. A brick circle, 18 inches in diameter, is built in the wall, to admit pipes to chimney without heating wood. Various other methods of running the pipe are used.

The bin or cooling room for a 16-foot kiln is about 16x20 feet. If adjoining, it is five feet lower than the kiln floor, with a doorway five by four feet, in halves, to put the hops through when dried; in this case, allow one or two feet of cold air space between the walls. Many build the coolers at a distance, connecting by trestle work (Fig. 120), as at Pleasanton, to reduce fire risk and cheapen insurance. The bin should be partitioned off into several rooms, so that not over three or four days' drying need be crowded into one room, as by this means the press in the room below (Fig. 121) can be started sooner.

The kiln floor is usually reached by a driving gangway for team and wagon, to a platform with a good shed over it, in which hops are deposited direct from the yard, until ready for loading the hop kiln. Wagons then pass down a gangway at the other end of the platform to the field level. Hence hop kilns are often built in a hollow to save as much hauling up a gangway as possible. A large elevator to carry the sacked hops from wagon to kiln floor is cheaper, and on some accounts better, where one has the power to run it.

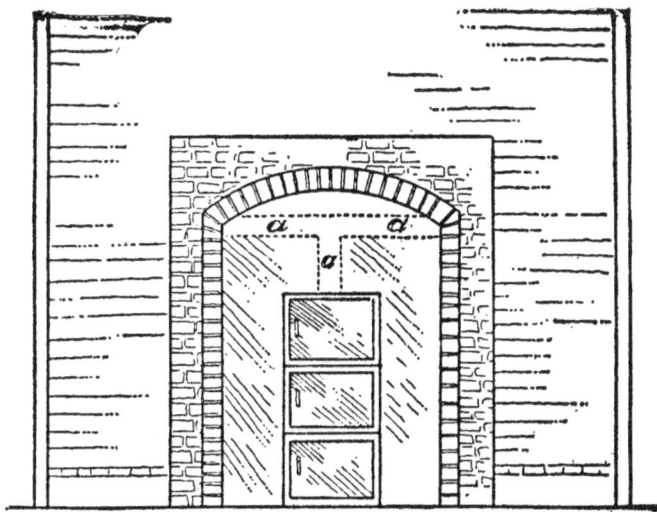

FIG. 104. STOVE FOR KILN, FRONT VIEW.

CHAPTER XIII.
Curing, Cooling and Baling Hops

FOR fuel, charcoal is used in Germany. Its fumes appear to have a beneficial effect on the hops, while its heat is intense, quick, and easily regulated. The German hop market will use no other fuel. In England, anthracite coal is employed, but coke is put on to keep the fires going, and some think it tends to impart the desired softness to hops. In America, dry wood is almost the only fuel in hop kilns.

One wagon and team can keep a ten-acre yard supplied with boxes and remove the boxes of hops to the kiln platform. Two men are necessary, and these will assist the dryer to load the kiln when ready, as it requires three men to load. The dryer and a fireman are required to attend the stove and drying, working alternately in shifts of twelve hours, changing at noon and midnight, so that each may have sleep in the night.

CURING THE HOPS

Everything being in readiness, the hops are delivered at the kiln loosely in large sacks, if picked in baskets, or in 120 bushel hop boxes. The floor cloth is carefully stretched—10½-ounce burlap or a strong duck is used for the carpet or kiln cloth; eight-ounce cloth is too thick and causes too much of the lupulin to fall on the pipes. The men wear sewed shoes, that no nails may tear the carpet. The sacks of hops are carried into the kiln and placed on the floor near where they are to be emptied, without dragging them across the carpet, and are emptied as lightly as possible, without shaking the floor, so as not to break the hops nor settle those already emptied. As fast as emptied, the hops should be loosened and leveled with a wooden barley fork. The floor can be laid to a depth of three to four feet, but at that depth it will require a long time to dry, and the bottom hops would be scorched while the top ones would hardly be dried. It is best not to lay them deeper than can be dried and moved in twenty-four hours, and the picking should be stopped when enough have been secured for this purpose. Therefore, it is bad policy to have too many pickers, as they become dissatisfied if compelled to lay idle any portion of the day.

FIG. 105. A HOP BALING PRESS.

Hops that have been heated in the sack while waiting to go on the kiln, will become smudged and discolored, and it is absolutely impossible to make a choice hop out of them, as nothing can be done to bring them back to their original state after they have once become heated and spoiled. There is no reason why hops should heat in the sack if growers would observe a little diligence. The heating of hops in the sacks is caused by either packing them in too tightly or from permitting the pickers to sit on them as if the sacks were cushioned chairs. Also avoid allowing the sacks to remain on the platform too long, and when there see that the sacks are not piled on top of one another. Do not pack too many hops in a sack.

FIG. 106. ELEVATION OF HOP HOUSE, NEW YORK—See Page 163.

On a deep "floor" the hops may have to be turned, or they may be scorched or imperfectly dried. Many careful hop men oppose deep floors and turning of hops, though practiced everywhere. The deep floor also requires excessive heat. If the hops get "packed" they must be stirred, using a long-pronged fork, with ends of tines turned up to avoid pricking the cloth. In New York, the floor is usually 12 to 18 inches deep, deepening as the harvest proceeds and the hops get dryer. A fan blast is often used to force a current of heated air through deep hops, and this may prevent the need of turning. On the Pacific coast, hops are seldom laid over 24 inches deep, and 20 inches are ample.

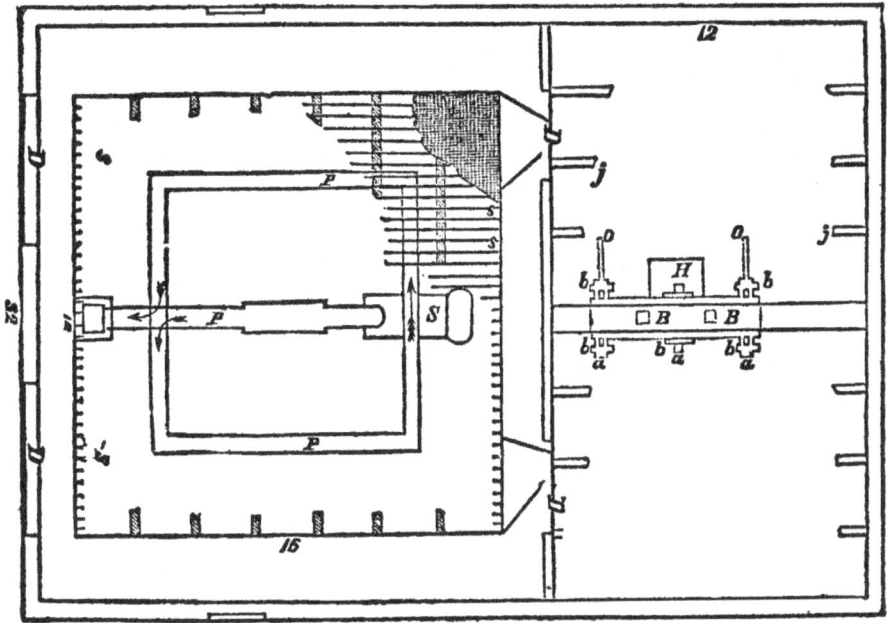

FIG. 107. GROUND PLAN OF HOP HOUSE SHOWN IN FIGURE 106.
Showing the arrangement of stove and press room. *S*, stove; *P*, pipe; *H*, trap door in room above to let down hops to press; *B*, *B*, bottom of press; *b*, *b*, keys to hold the sideposts of press; *o*, *o*, railway for moving press under the hole *H*. In this figure, the positions of the joists *j*, *j*, and slats *s*, *s*, of the floor above are also shown.

The object in curing hops is to evaporate their excess of water without loss of other qualities, and in the least possible time. Green hops contain from 70 to 75 per cent. of water; cured hops from 7 to 10 per cent. This change is usually effected in 12 hours, the morning pick going in at noon and the afternoon harvest at midnight. As a floor four feet deep and 16 feet square will contain about 45 boxes, or 900 bushels, the green weight of 4,500 pounds shrinks to some 1,100 pounds. Thus, the 3,400 pounds of water in the hops has to be evaporated and carried off during the 12 hours. Hot air to evaporate the water, and a strong current to carry the vapor off are essential. Hence, the need of an abundant inlet of cool air into the stove or heating pipes, and of ample ventilation to draw off the hot vapor in such a way as to create a strong draft or current of heated air through the light and fluffy mass of hops.

173

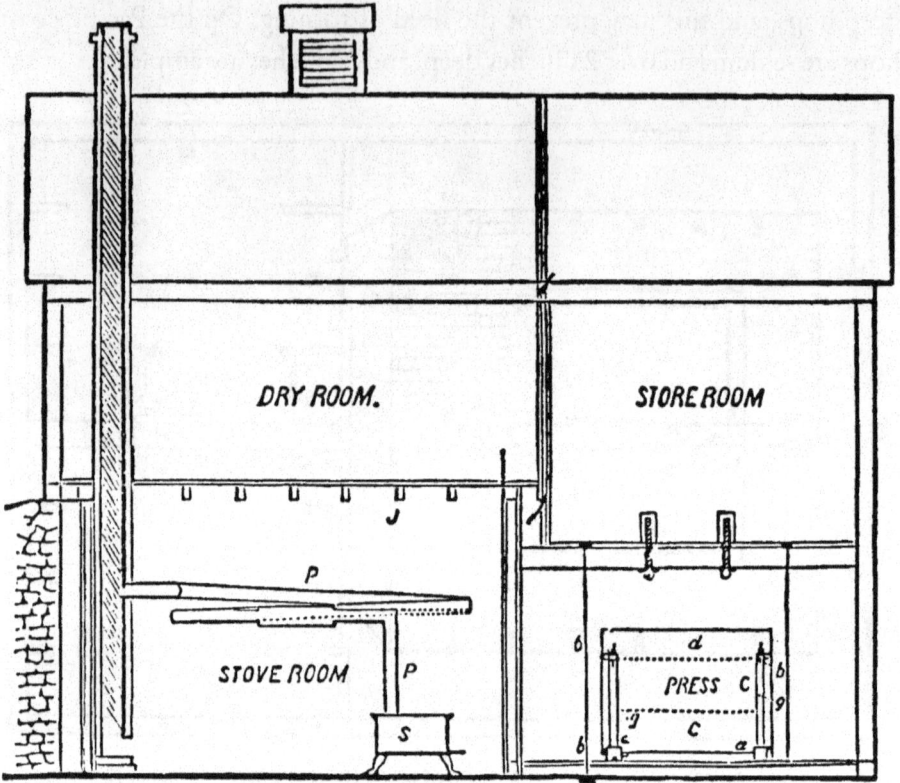

FIG. 108. SECTION OF HOP HOUSE SHOWN IN FIG. 106

Showing stove, dry, store and press rooms. *S,* Stove; *P,* pipe: *C,* movable sides of press; *d,* upper plank of press; *g, g,* posts to support sides of press; *b, b,* iron rods, which connect the bed-sill with the strong beam above.

After the floor is laid, the fire should be started and the heat raised to the desired point in two to four hours. If the heat is rushed up quickly, it will cause the hops to pack, whereas they should be kept so light that the heated air will freely circulate about every hop. At the start, open wide the ventilators in the cowl, to let the steam off freely, and as long as steam is emitted, see that the sulphurous acid fumes permeate the air. When the steam is gone, or nearly so, stop the sulphur and close the ventilators halfway. This is in about 10 hours, generally, on a 12-hour cure, depending on the condition of the hops. Finally, close the ventilators tight, to allow the top layers to be dried off.

It is not possible to describe in words the condition hops are in when the cure is done and the floor ready to renew. It must be learned by actual experience, just as the qualities and curing of cigar leaf tobacco can be judged only by experts. Meeker attempted to do this in his book, from which we quote:

"An ideally cured hop would show only a wilted stem, or core, of a purplish-green cast, being soft to the touch and flexible; the globules of lupulin, standing out prominently, bright and unchanged from that of an uncured hop. In practice, however, most of the stems are not only wilted, but are dried so as to be brittle and harsh to the touch, and show the sharp corners, which will be readily understood by any one taking a specimen between the thumb and finger and rubbing the hop to pieces. Because of the presence of these over-dried hops, we are able to turn off the flooring with a small percentage of fat hops, being those whose stems are not wilted, but show as green as when placed in the kiln. Floorings, with 10 per cent. of such stems, may be turned off, and yet keep, if otherwise well and evenly cured and properly handled afterward, though I should by no means advise leaving so large a percentage; probably not five per cent. of such stems are left in ordinary practice. Whatever there may be will have disappeared in a couple of days, and such stems as were green will be wilted and the moisture absorbed by the balance of the hops. The after-handling consists in forking the hops over after they have lain a few days and have begun to warm up, as hops in bulk will do where not dried thoroughly. A better plan is to cure at a lower temperature, which will insure a larger percentage of wilted stems, and less of those with sharp corners, as likewise of the green stems and a more even color."

If hops are slack-dried, they will "give" when cooled off. If over-dried, they will fall to pieces or shell badly, feel harsh and the stems will be brittle. This last state can be helped by putting a quart of salt in a pan on the stove and shutting the ventilators for a short time—a little trick that will soften and toughen the otherwise brittle hops. Even on brittle hops, salt must be used with care, and never on hops that are all right. It causes hops to absorb moisture, especially in Oregon and Washington. A hop that absorbs moisture before being baled is liable to be clammy and boardy.

When the hops are done, draw fire at once, and open ventilators, and allow the heat to go off; then carefully remove the floor into the cooling room. This is done with a rake or box shovel (Fig. 109).

175

COOLING AND BALING

Before removing to the cooling bin, the floor of hops is allowed to cool off for an hour. Where the box shovels (Fig. 109) are not used, hops are taken from the floor in wheel scoops, operated by one or two men, or pushed by rakes into cars (Fig. 112). The floor-cloth is carefully swept with a peculiar broom (Fig. 109), to save all the lupulin and dust, and as a guard against danger of fire. If a floor comes off red, discolored, or with traces of mold, keep such hops separate; don't mix, or you will reduce the price of all the hops. Strive by all means to keep qualities separate, if more than one, so that the buyer gets his goods as he bought them, and thus insure a reputation for yourself.

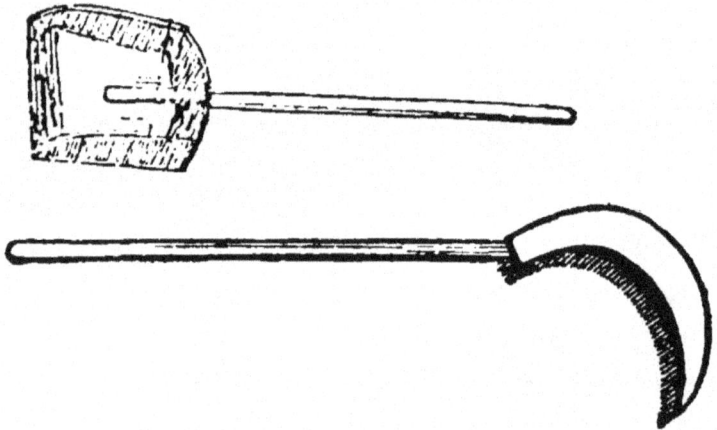

FIG. 109. SHOVEL AND BRUSH FOR USE IN KILN.

English and German practice is to pack the hops, while still warm, into sacks about six by three feet, containing about 125 pounds and called "pockets." These bales are usually sent to market at once, and if not promptly sold to the brewer (who puts them into cold storage) are stored in large, cool, airy warehouses, so stocked as to permit a free circulation of air about each bale; otherwise, the hops may become crusted and damp. Prompt packing while warm prevents the escape of the volatile sulphurous acid gas, the retention of which in the bales adds to, or rather preserves, the brewing qualities. This gas is exceedingly volatile, and the more it escapes before hops are packed, the less will be its beneficial effect. Kammerer showed in his tests

176

at Nuremberg that sulphured hops left open and unbaled for four weeks, steadily lost their binoxide of sulphur, until after four weeks they contained only 25 per cent. as much as when first off the cloth. Another argument for packing while still hot, is that the hops contain less water than in any other period, and if allowed to cool, will rapidly absorb water, thus partly counteracting the object of the cure. Scientific experiments on all these points would be highly interesting.

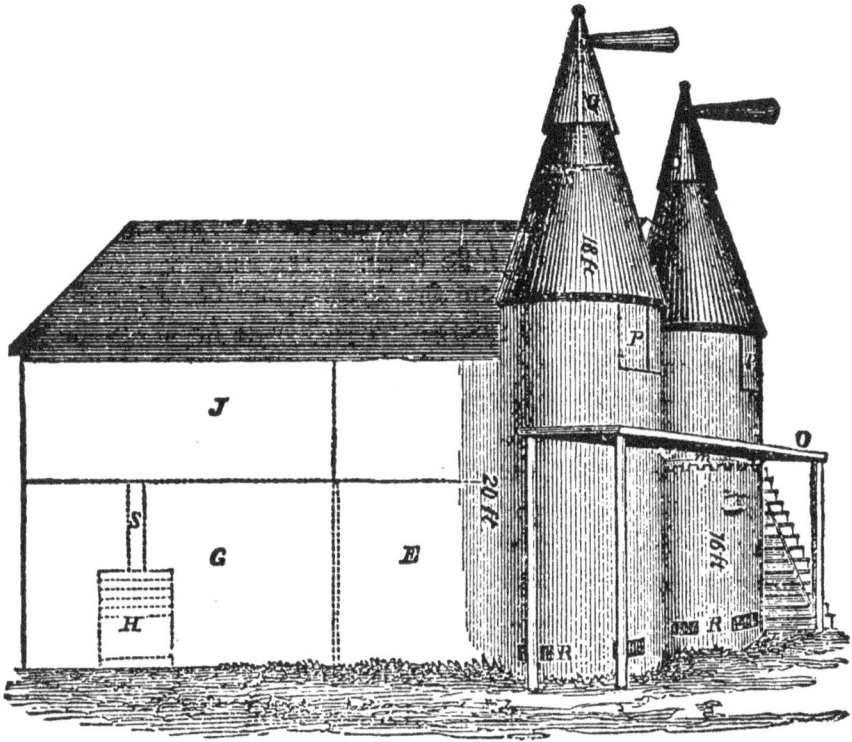

FIG. 110. ELEVATION OF DOUBLE KILN, NEW YORK.

In America, however, thorough cooling is the rule, partly because when baled cold the hops are alleged to keep much better during the often long interim before they reach market. In some cases, they lie in the bin twenty-four hours, and are then put into another store room for ten days to sweat. The dried hops remain in the bin until they commence to toughen or "come in case," which takes from three to seven days, depending upon the temperature and density of the atmosphere. But if allowed to lie too long, they again become very brittle and break badly in pressing; if left until they again show moisture, they pack in the press hard and solid, and samples taken from them are what dealers call boardy and which they claim are slack-dried. If hops are to be held by the grower for some months, there is considerable testimony to show that they can be kept with less injury in bulk than in bales. Meeker inquired particularly into this point, and still believes it is fully demonstrated. Of course the warehouse must be kept cool and very dry, so the hops in bulk will neither heat nor absorb moisture.

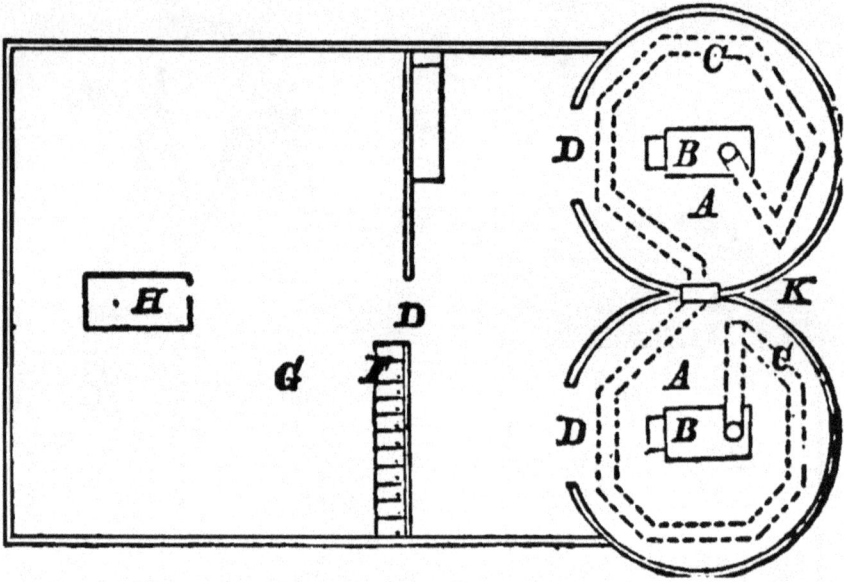

FIG. 111. GROUND PLAN OF DOUBLE KILN SHOWN IN FIG. 110.

Meeker covers another important point as follows: "One objection to the practice of baling immediately after curing is that the grower's crop will not run as even in quality as if carefully stored, in order that they may mix the whole thoroughly. The earlier picking will be lighter; that is, not so rich as the later, besides no field of hops will be of exactly the same quality and color, even if picked on the same day. To most effectually mix, so as to have the whole crop uniform, the warehouse should be filled in layers, first covering the whole floor about two feet deep, and gradually fill by adding successive layers; then when baling, by taking the whole depth of the pile there will be no appreciable difference in color or value. This plan gives uniform samples from every bale, a point highly desired by the buyer."

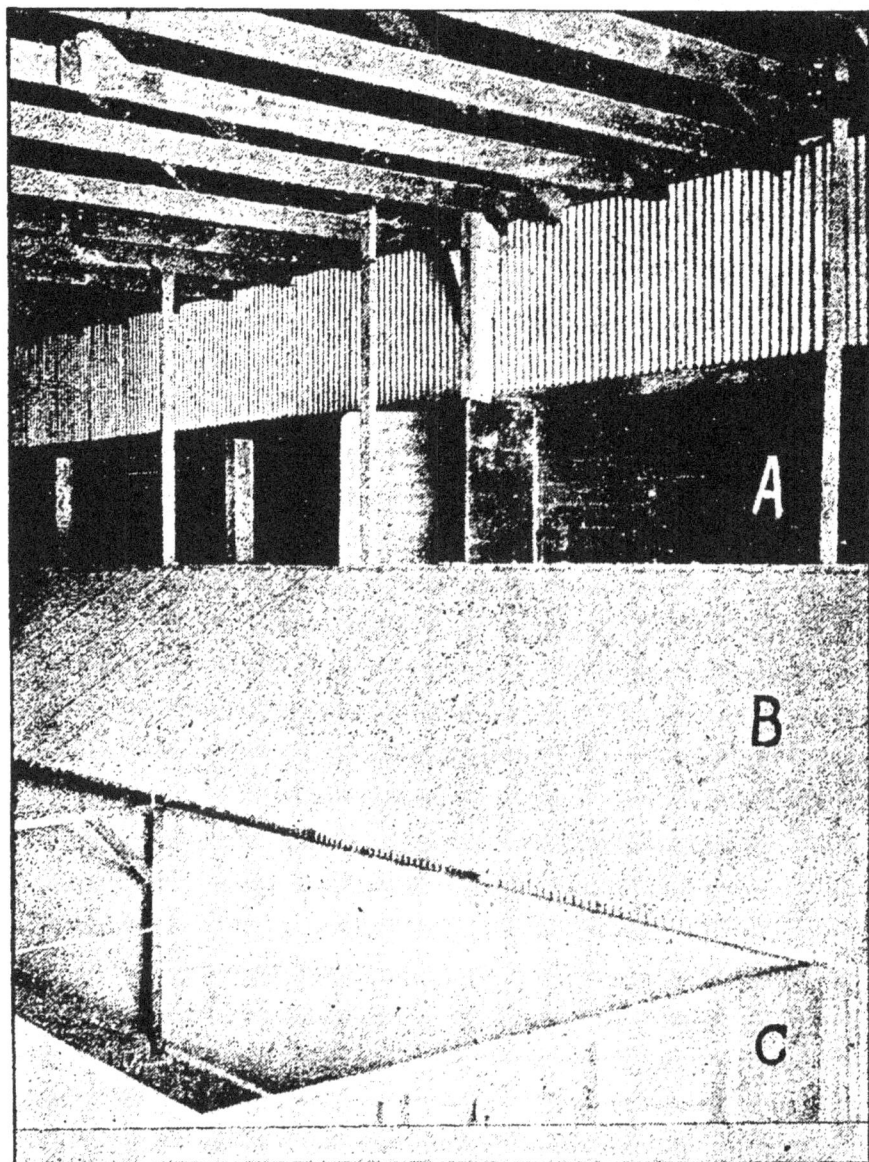

FIG. 112. CAR AT KILN TO RECEIVE HOPS (Pleasanton).

A, Upward sliding doors of kilns, through which hops are pushed from floor; *B,* wooden apron down which hops slide (to prevent their breaking), with the car, *C,* ready to receive the dried hops to be carried to the coolers.

Great care should be exercised, so as not to break the hops during the process of baling. Many growers tramp the hops with their feet without using any board to rest on the hops. In fact, we know cases where growers actually stamp the hops in the presses with their feet. A horse-power press that does not require any tramping is, of course, preferable, but these presses are expensive and not all growers can afford them. A hand-power press that requires the follower to be run down more than once, can be made to bale hops in perfect shape by using the board mentioned above, and a springy motion of the knees, while standing in the press, instead of tramping them, will prevent the hops from breaking. A hop that is broken from any cause, whether from baling or otherwise, is far from a "choice" hop. It may be all right in other respects, but the mere fact that it is broken will detract from its selling, as well as its brewing, qualities, and, in addition thereto, a broken hop will naturally age more quickly than a whole-berried or a "flaky" one.

FIG. 113. THE HARRIS HOP PRESS.

Before beginning to pack the hops in bales, get ready the sacking (weight, 20 to 24 ounces per yard), twine for stitching, brush and stencil brad, with four men to do the work. Cut the cloth four inches longer than the press, thus allowing two inches on each side for stitching. Put the bottom cloth in, fill up the press to the middle with hops; then let two men get in to press them down with their feet, having a three-fourths-inch board, covered with cloth, the full size of inside of press (less one-half inch all around), for the men to stand upon so as not to break the hops. Then get out, remove the board, again fill up and again press the hops down with the board. Next fill up with hops to the top, lower the press and squeeze down; lift up, fill up again, insert top cloth cut just the same as the bottom, and press down with the clamps. Open the sides and again press down about two inches more. Sew on the cloth sides with a sacking needle and twine, and then draw. Deliver the bale to the fourth man, who will complete the sewing of the sides and store away. The press should be run down until the cloth will lap at least one inch on each side of the bale, and the cloth should be evenly but not too tightly drawn, and sewed with short, even, lock stitches, causing the strain when the bale is loosened to come evenly on all the stitches. After being removed from the press, the ends should be sewn in at once before the bale commences to sweat. A number of styles of presses are used. Large plants employ power presses, which do away with tramping the hops and save breaking them. The two presses at Pleasanton can each turn out 80 to 100 bales a day, extraordinary capacity being necessary to handle its crop of over 3,000 bales.

FIG. 114. THE PRESS WITH FRONT REMOVED.

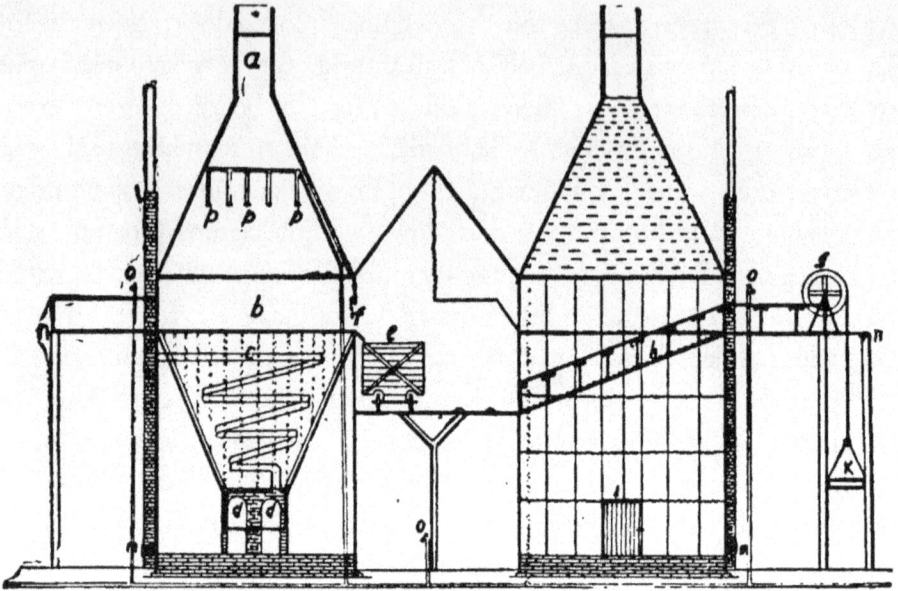

FIG. 115. PLEASANTON HOP KILN. FRONT ELEVATION. INTERIOR
VIEW GIVEN OF KILN AT THE LEFT.

a, Ventilation regulator (opens and closes by pulley ropes extending to kiln floor);
b, kiln floor 30x30, built of 1½xl inch slats set on edge, with 1½ inch space between
each slat to allow heat from furnace and pipes to pass through hops. Over the slats
is laid a carpet of 10-ounce burlap to prevent hops rom falling through. On this
floor the green hops are spread for drying process; *c*, section of heating pipes
detailed in Fig. 118; *d*, furnace, of boiler iron, 6 feet long, 4 feet diameter, with brick
supports and brick enclosure provided with draft doors as shown; *e*, car with
movable sides and bottom, used to transport the dried hops from the drying floor
to coolers (Figs. 120 and 121); *f*, upward sliding doors through which the dried
hops are shoved from kilns to cars: *g*, elevator wheel, for hoisting the green hops in
sacks from wagons to kiln floor. About a ton are hoisted at one time, the hops being
placed on *k*, the elevator platform; *h*, stairway built outside of kilns and connecting
kiln platform with car track; *i*, door to furnace room; *j*, car track built on trestle, 20
feet high, connecting kilns and coolers; *l*, brick and iron chimney 50 feet high,
provided with, *m*, dampers for regulating draft in chimney; *n*, elevated platform
outside of kiln room; *o*, 2½ inch hydrants connected with 5-inch water main from
tank and steam pump; *p*, 1 inch automatic sprinklers, 5 over each kiln floor, and
connected with same water service.

Bales should weigh not less than 185 pounds nor more than 200 pounds,
as near to 185 pounds as possible. To make the bales conform to this limit,

it is necessary to weigh each bale as it comes from the press, to be able to determine how to fill the press for the next bales; or it can be closely judged by noticing how many "scoops" are required per bale. Stencil the bale plainly with your brand, which should also give grower's name, post office, county and state. Don't stencil the weight. This will be done in the buyer's presence. He will deduct five pounds from each bale for the cloth (which is the law in New York state), and unless this is stipulated, he will want to deduct seven pounds.

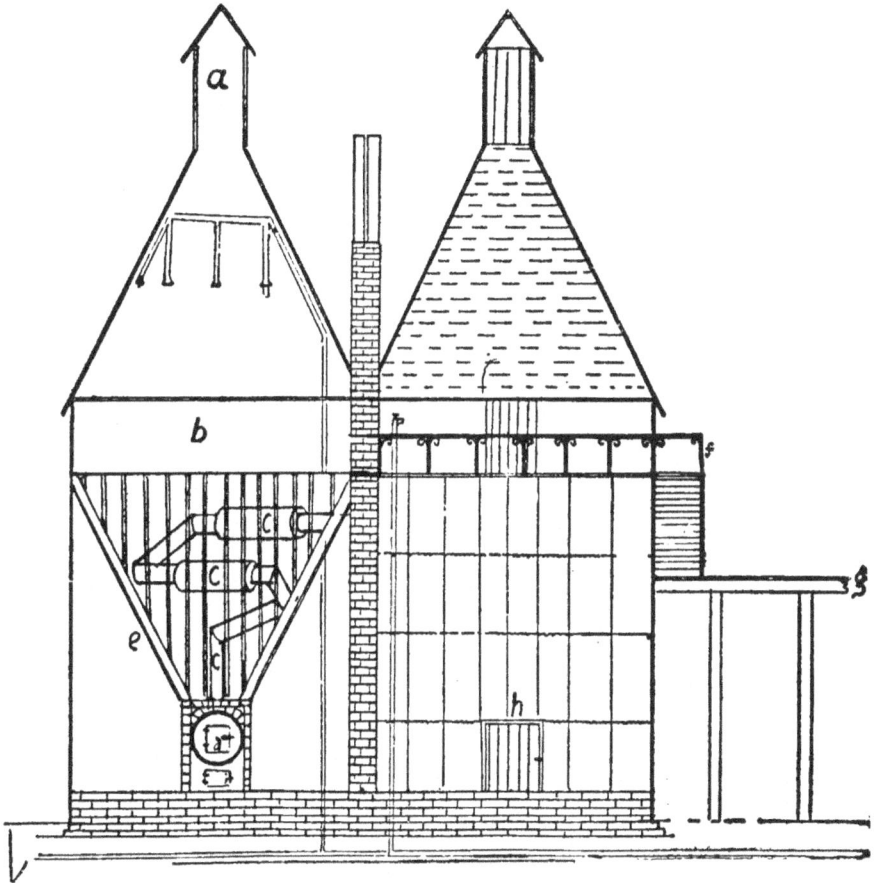

FIG. 116. SIDE ELEVATION OF KILNS—INTERIOR VIEW AT LEFT.
a, Ventilators; *b*, hop drying floor; *c*, heating pipes; *d*, front of stove showing brick enclosure; *e*, sides of hopper, detailed in Fig. 117, *f*, elevated platform; *g*, trestle supporting car track; *h*, door to furnace room; *i*, door to hop drying floor.

In storing hops, the bales should be set on end, not touching each other, and if they are to be stored more than one bale deep, a couple of boards can be laid on one row for the upper row to stand on. If they are to remain in storage any length of time, the bales are best turned the other end up every 10 days or two weeks.

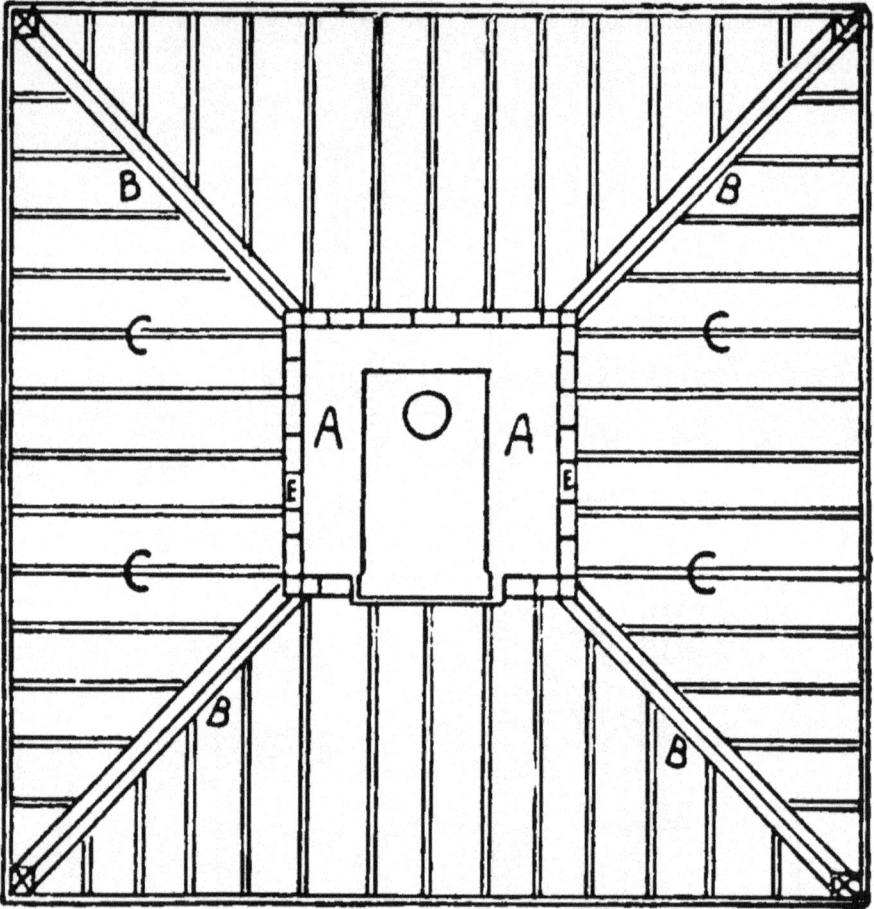

FIG. 117. DETAIL OF HOPPER.

A, A, Space occupied by furnace (shown at *d, d,* in Fig. 115) with surrounding wall. On this wall rest the main supports, *B,* of hopper, and smaller supports, *C.* To these supports are nailed iron laths (covered with plaster), making the structure practically fireproof.

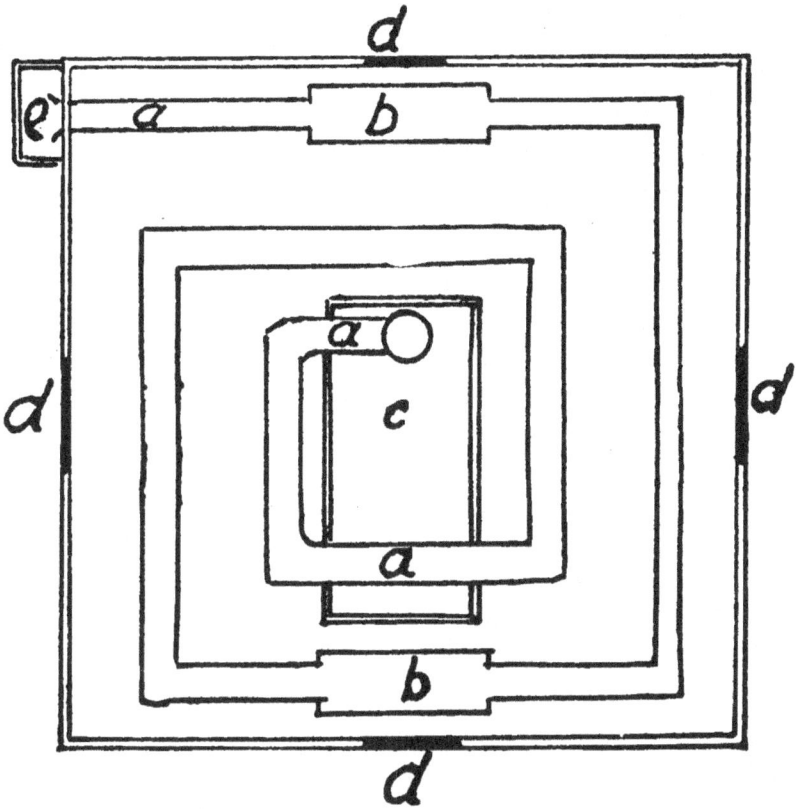

FIG. 118. ARRANGEMENT OF HEATING PIPES AND DRUMS.

a, a, 12-inch main pipe leading from furnace and with continuous turns and ascending, as shown in Fig. 116, finally ending at chimney, "*e.*" At intervals 22 inch drums (*d*), 8 feet long, are placed to give more heating surface and to more evenly distribute the heat. The pipes are of sheet iron with riveted seams.

ADDITIONAL NOTES ON CURING

Hops from young yards, and hops damaged by mold, require to be entirely dry, and may be finished up with a few degrees more heat. They also require pressing sooner, as they slack and become moist sooner, on account of the core or stem being larger.

Green hops are liable to become heated if allowed to remain in bulk, even over night, and it is advisable to stir them late in the evening if they have to be kept over. Heating will cause the lupulin to drip out. If the hops should come

185

to the kiln hot and wilted, it will be found to be a good plan to thoroughly dampen them with a sprinkling pot of water as the flooring is being laid. This causes them to freshen up, and the escaping steam will open up the leaves of the strobile, thereby letting the heat season out the core of the hops without baking all the life out of them by exposing them to long continued heat. If the hops are to remain long on the platform before going into the kiln, the sacks should be set so as not to touch each other, the mouths of the sacks opened, and the person caring for them should run his hand and arm down through the center to the bottom of each sack, then grasp a large handful of hops and draw the hand out. This will loosen the hops up and leave a hole through the center, permitting the air to circulate freely, preventing the hops heating in the sacks.

The proper temperature for curing hops is a matter of dispute. The trouble is that with limited kiln accommodation, it is necessary to cure the hops in 12 hours, and to do this extreme heat is required. The kilns can hardly be used for other purposes; they are costly to build and maintain, and the average planter finds it more profitable to run the risk of a quick cure than to have sufficient accommodation to cure his hops at a lower temperature, which would require 24 hours to each load. The importance of the slower cure at a lower temperature is becoming recognized, however, and at Sonoma and Pleasanton only one floor is placed on the kiln every 24 hours, and the hops are cured at the lower heat. Even then, the cured hops hardly compare with the sun and air dried hops of Germany.

The usual temperature for a 12-hour cure in this country is 140 degrees F., as tested by a thermometer hanging in the midst of the hop floor or immediately over it—not at one side; formerly it was 160 to 180 degrees. In England, the temperature ranges from 120 to 140. Where the 24-hour cure is followed in California, the heat is kept at 100 or 110. Whatever the temperature decided upon, it should be kept uniform.

Should the heat go up suddenly, open the doors of the kiln and the ventilators at the bottom of the house, so as to reduce the temperature to the desired figure. The utmost care and constant attention, combined with good judgment, are needed to preserve the temperature, to watch the hops and to see that the cure proceeds properly.

FIG. 119. PLEASANTON KILNS AND COOLERS.
Car full of cured hops is being run into the cooling house over the trestle. Showing also some of the employees.

The extreme heat absolutely destroys a considerable proportion of the essential oils and other brewing qualities. Mr. Meeker was perhaps the first American writer to call attention to this point, which he did in 1883, and what he then wrote in his book, "Hop Culture" (long since out of print), is equally true to-day:

"This substance is most sensitive to injury by high heat, and hundreds of tons of hops are injured annually, and in many cases their value almost totally destroyed by the careless or ignorant manner in which they are dried. The writer knows by actual experience that when hops are subjected to a heat of over 160 degrees Fahrenheit, there is visible to the naked eye, a change in the appearance of the lupulin in many samples that can be selected in a flooring of hops, though not all will show the effect alike. We are led to believe from this that either the length of time after being dried that the hops are subjected to the current of heated air, or else some unknown condition of the hops before going on the kiln, governs this visible sensibility to heat; be that as it may, the fact stands out prominently so that any observing hop grower can demonstrate it with no expense and but little trouble. As the heat is increased the change becomes more apparent, until at about 180 degrees the globules begin to disappear and run together, presenting a dull brown or red appearance, of all shades, according to the degrees of intensity, and, as we believe, duration of time the hop has been subjected to this high heat. If to the naked eye there is a visible change in this delicate substance, from the effects of heat, how much more apparent it becomes when subjected to the rigid scrutiny of the chemist or the practical test of the brewer. The extract or bitter principle of the hop, according to Thausing, assumes a reddish-yellow color when heated above 140 degrees Fahrenheit, and when cooled off can be rubbed into a fine powder... (At 212 degrees F. the hop bitter swells up under decomposition, and combustion takes place, with a sooty flame.) And yet, an American authority recommends 180 degrees F. as a safe temperature. We know from experience that it fairly cooks the hops and destroys much of their value."

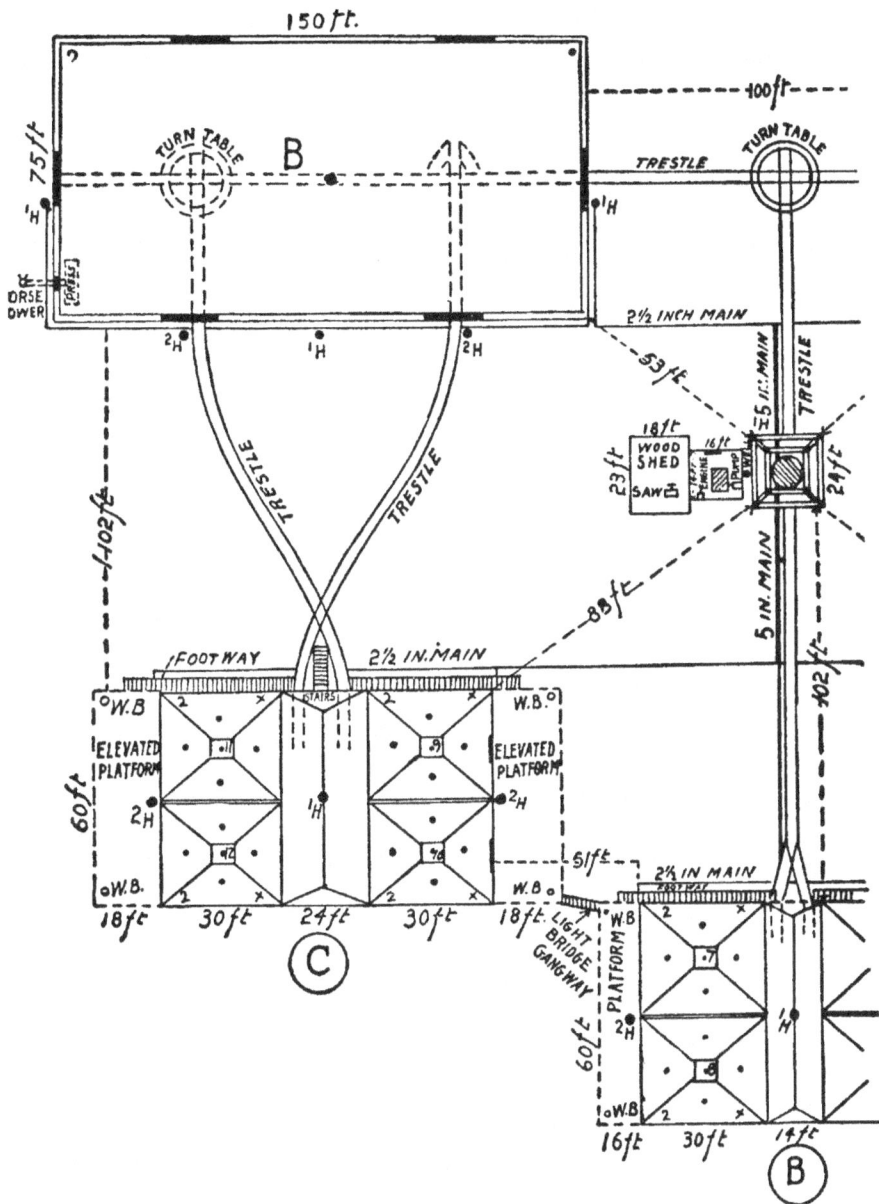

FIG. 120. GROUND PLAN OF THE PLEASANTON ESTABLISHMENT.

Left half of plant. It is duplicated on the right side, not shown hereon. *C*, Are the double kilns shown in Figs. 115 and 116; *B* are similar kilns, and another pair of kilns and cooling houses are at the right in the duplicate half of plan not shown above. *B*, Is the large cooler or warehouse. The tower for water tax is at the center, with engine, pump, etc., nearby.

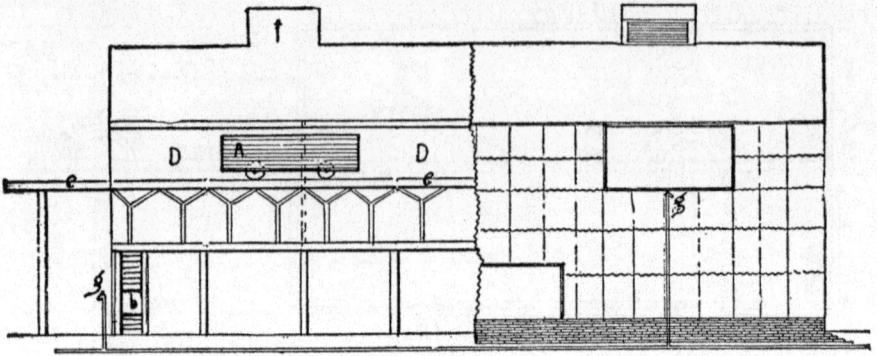

FIG. 121. COOLING HOUSE FOR HOPS—SIDE ELEVATION.
(See also fig. 122.)

The Pleasanton company has two of these coolers, each 150 ft. long by 75 ft. wide, of corrugated iron, with inside double wall of wood. Second floor of each cooler is divided into six bins, *D D*, three on each side of house, into which the dried hops are dumped from car, *A*. Trestle and car track extends the length of the buildings and is provided with turntables, enabling the cars to be run from any of the kilns to any of the twelve bins. Ground floor used for storing baled hops. *B*, Hop presses operated by horse power. *G*, Two and one-half-inch hydrants connected with five-inch water main from tank.

Whitehead, writing in 1897, confirms his earlier statements in 1893, and in still earlier years, to the effect that "the merciless treatment of stewing or baking, to which English hops are subjected, causes an absolute, visible loss of lupulin, besides the loss of ethereal essences." English hops, dried slowly at a temperature never rising above 100 degrees F., were found on analysis to contain larger quantities of resin, oil and bitter principles, and at the same time considerably less moisture, than Spält hops cured by the same process. Worcester hops dried in this gradual manner were found to be uniformly rich in desirable qualities and to have far less moisture than the best brands from other hop-producing regions of the continent or America cured in the ordinary way.

Meeker pointed out that there is also a greater loss in the non-keeping qualities of high-dried hops than those cured at a low temperature. It would be interesting to go into more details on this important subject of the proper temperature, and to discuss its scientific aspects, but the facts are as stated, and this is sufficient for all practical purposes.

190

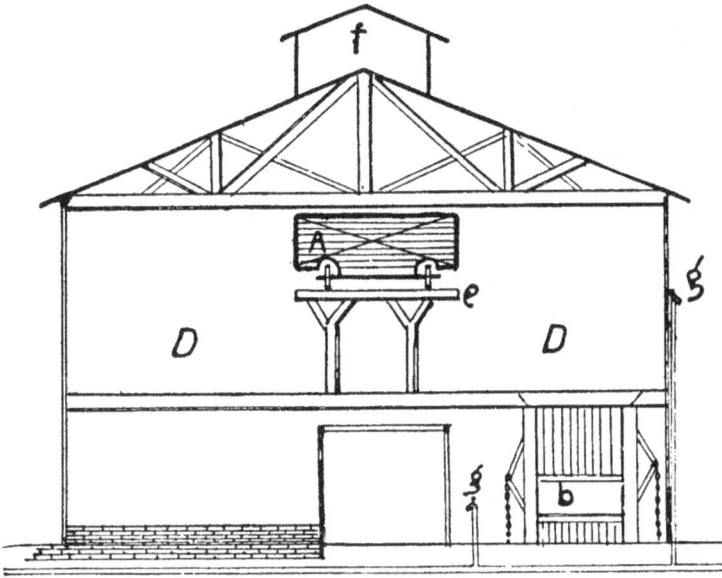

FIG. 122. COOLER—END ELEVATION.
See Fig. 121. For explanations.

THE SULPHURING OF HOPS

This is done in several ways. In German curing kilns, also in English oast houses, the sulphur is thrown directly upon the fire, usually after the hops have begun to steam, another dose of sulphur being put on the fire when the hops are turned. From 10 to 20 pounds of brimstone are burned for each floor on a 20-foot square or 16-ft circular. In New York, the hop stoves have a flange on which to burn the sulphur, but as it burns too rapidly, the custom there (and also on the coast) is to put the sulphur in a pan on the ground, near the stove, and set fire to it with a few hot coals or with a red-hot iron.

If the hops are nice and free from rust, one pound of sulphur is used for a floor in a 24-foot kiln, but when very rusty, from two to five pounds are used; others use only two to three ounces at a time, and burn it three times, first when the hops are warmed one-third the way through, and last when the heat has reached the surface. On the coast, from one to four pounds of sulphur per 100 pounds of dried hops is the range. Usually a little sulphur is burned, with ventilators partly closed, just before the hops are done, to finish off the drying.

191

The bleaching effect is not as much at this time as when the hops are more moist.

FIG. 123. CIRCULAR KILN, MONTGOMERY COUNTY, N. Y.
See ground plans, Figs. 99 and 100.

Some Oregon growers find that "the best way to burn sulphur is in iron kettles, hung by a hook on wires stretched across each corner of the kiln, and high enough to be above the heads of anyone passing underneath. These kettles can be lifted off with a forked stick and set on the stove to get hot when the fire is started, and again hung on the wire as soon as the sulphur is burning well. The kettles can be replenished by dropping sticks of sulphur in them while burning. They should be kept burning until the hops are dry enough to rattle on top. The amount thus used will be found to be about one pound to each 20 pounds or 25 pounds of dried hops." Meeker burns the sulphur outdoors, but close to the fan, by which the fumes are sucked up with the air and forced in a powerful current through the hops after the air is heated.

FIG. 124. IMPROVED KILNS AND WAREHOUSE OF IRON.

These buildings being absolutely fire-proof, no insurance is carried on their contents. See Page 218

From photograph of one of Horst Brothers' Hop Ranches.

No other chemicals should be used, as they destroy the vitality of the dried hops and cause them to fall to pieces and look and feel husky, like chaff. There are four reasons for using sulphur: 1. The fumes decrease the hygroscopic power of hops; that is, render them less able to retain the moisture that is both within and without them, and thus the fumes help to carry off the water in the hops. This is a vital point in favor of sulphuring. 2. The fumes bleach the hops, dispelling spots, dark colors, etc., and making the hops brilliant, glistening and attractive in appearance and feeling. 3. The preserving effect of the fumes prevents deleterious transformation of the albuminous and other principles in hops, the proper preservation of which is so essential. 4. The antiseptic action of sulphurous acid fumes kills all fungus germs, such as mildew, and is a partial safeguard against infection by mold and other fungus growth.

A few years ago brewers claimed that sulphuring hops in drying added weight to them, and they persuaded many growers to dry them without sulphur, but when they got the hops they were badly disappointed. The hops contained moisture and impurities and lost strength rapidly, and the next year all the brewers' chemists advised the liberal use of sulphur. Much has been said against the sulphuring of hops, but the weight of scientific and practical knowledge, and the experience of brewers, is overwhelmingly in favor of bleaching with sulphur fumes when the process is properly conducted. The Bavarian government's commission, under the chairmanship of the great Liebig, made a most elaborate investigation as long ago as 1855, and established the great truth that when this process is properly done nothing but good results. All scientific and practical experience since has abundantly confirmed this result.

But too strong a sulphurous acid gas, made by having too much sulphur or by not allowing proper ventilation, is certainly bad for the hops. Too much gas, like too much heat, tends to give hops unfavorable characteristics. Either or both also remove more water than necessary. The grower's object should be to retain as much water as can be done without detriment to the keeping qualities of the hops, thus having a greater weight for market. American and English hops in market usually contain fully 7 to 10 per cent. of water; they might contain 10 to 15 per cent. without much danger of heating. Re-sulphured hops in Germany contain 20 to 22 per cent. of water.

The use of impure sulphur is even worse. Only the very best virgin roll brimstone should be burned. A few cents saved in buying cheap or inferior sulphur may cause the loss of many dollars in selling value of a crop.

The prejudice against sulphured hops, in the minds of certain brewers, is due to the fact that the process is unmercifully abused by irresponsible persons. Old hops, that have become moldy or discolored, and new hops of inferior quality, are often given a beautiful appearance by re-bleaching, which makes it difficult to judge their merits. This trick is much resorted to in the mixing of poor hops and inferior growths with those of better qualities, so as to palm off the mixture as composed wholly of the latter. This is practiced to such an extent by German dealers that England buys German hops only when forced to do so.

FIG. 125. A GLIMPSE OF KENTISH OAST HOUSES.

CHAPTER XIV.
GRADING AND MARKETING HOPS

FIVE different grades are recognized in all hop markets of this country and England, both by the buyers and brewers, and they are classified as follows: Fancy, choice, prime, medium, poor. It takes considerable practical experience to be able to determine the quality of a hop and place it in its proper grade. There is no means of testing, and one is compelled to use his judgment, as well as the senses of seeing, smelling and feeling, as the appearance, flavor and feel of a hop are all essential features to determine its brewing qualities. We have been unable to find a single buyer of hops in either America, England or Europe, who employs chemical analysis as a basis for judging the quality or value of hops. In fact, it is generally recognized that the correct characteristics of hops cannot be quantitatively measured, but are a matter of judgment. Now, because judgment of quality may differ so much, and because such differences do prevail, we have taken great pains to get the views of leading growers, dealers, merchants and brewers in the United States, Canada, England and Europe upon this delicate subject. The concensus of opinion seems to agree in a general way upon the following as a fair statement of principles.

Hops in the trade are described and designated in three principal terms:

1. By their relative quality for a given country or section.
2. By the year of their production.
3. By the country or section where grown.

These three terms are connected as follows: "Choice 1898 Pacifics," representing as just stated, quality, year and locality. Under such designations a large proportion of the world's hop crop is bought and sold for immediate or future delivery, and of hops already grown and hops to be grown during the following year or term of years.

The quality is usually described in five so-called grades, namely: "Fancy," "Choice," "Prime," "Medium," and "Common." "Fancy" represents the very finest quality, or best selections from "Choice." "Choice" represents the first average quality, "Prime" the next lower average, then "Medium" and finally "Common," which is the lowest average quality.

All hops may be designated in the above grades, excepting such hops as are heated or are liable to heat, or rot, on account of insufficient drying. This class of hops is called unmerchantable and is handled on its merits—or rather its demerits—and is never bought or sold on grade.

The year of production is described by using either the year in figures, or using "new crop," or "old crop," or "old olds," as the case may be.

The country or section is designated as, "American," "English," "Germans," "Belgiums," etc., or by any of their respective subdivisions. These subdivisions may be carried down as far as one pleases, say to the state, county, town or other described district. As illustrating this it may be said that a hop grown near Healdsburg, California, would be known in the hop trade under any of the following localities, namely: "American," "Pacific Coast," "California," "Russian River," "Sonoma," or a Healdsburg hop, each successive title serving to narrow down the location where the hop is produced. Likewise in the London market the terms "English," "Kent," "Weald of Kent," and even the name of the particular plantation are used. In Germany such classification is carried to an almost absurd subdivision.

The placing of a given hop in its grade is largely a matter of the likes and dislikes of the expert, and the experts themselves rarely agree among themselves as to the proper grading of a line of hop samples. With the brewers the difference of opinion as to the grades is even greater than it is with the hop dealers or experts. Again, in naming a grade for a hop, even an expert is often in doubt whether to place it in one grade or in another; thus, if a sample looks a little too good for a given grade and not quite good enough for the grade next higher, an intermediate grade is often used, which is expressed by connecting these two grades with the word "to." Thus: A hop that one considers better than a prime and not quite equal to choice, would be expressed as "prime to choice," though these split grades are not so frequently employed as the general terms, fancy, choice, prime, medium, common. Split grades or gradations or peculiarities of quality are sometimes spoken of as "strictly fancy," "shipping grade," "good brewing quality," "trash," and perhaps a hundred other minor terms are sometimes used, taking the hop world as a whole, but fancy, choice, prime, medium, common, are the accepted standards.

FIG. 126. LOADING BAGS OF HOPS FOR THE KILN, CALIFORNIA.

There are some features of a hop that are liked or disliked by everyone, though in many other particulars what one purchaser prefers, another condemns. Brightness in characteristic color, strong, healthy flavor, freedom from disease or vermin damage, proper maturity, clean picking, flakiness, softness in texture, abundance of lupulin, are all conditions that are appreciated by everyone. Yet as to color, that is, whether green, yellow or red; or in flavor, whether from English, German or other varieties; or as to largeness or smallness of the berries, these are conditions the like or dislike of which is dependent entirely upon one's own fancies. Mold, sour or burnt smell, dirty picking, broken and powdered buds, lack of lupulin, no flavor and dullness of color are considered defects by everyone. Even as to color, there can manifestly be no arbitrary standard. Hops grown in a cloudy season in New York state, even with the most perfect curing, may come out a sort of slate color, instead of the usual bright golden yellow or greenish yellow characteristic of state hops in a favorable season; yet the hops of the darker hue may be so attractive in other respects as to command top prices. Again, hops of a greenish cast or that are "of the green order," as Wahlberg and the hop men say, may be choice in every particular and just the kind that takes best with certain brewers. Indeed, color in hops, as in art, seems to be largely a matter of taste and habit, some people wanting one shade, some another.

The chief contention is as to what constitutes a choice hop. No doubt this is accounted for to a large extent by the failure to distinguish between the use of the word as an adjective or as a noun. For example, in contracting for a growing crop, it is evident that the term is used as a noun, and in this sense it would signify a hop which is to be chosen or selected. In this case the hops might prove to be choice hops of a certain growth and yet not possess that perfection of color, freedom from mold or vermin damage or impurities, and that perfect cure which may be implied by the adjective "choice" or "fancy." To some this distinction may appear subtle, but an illustration will prove that it may be vital.[1]

---

[1] For this illustration we are indebted to Mr. Hugh F. Fox, a well-known American commission merchant in hops, formerly secretary of the hop trade committee of the New York produce exchange, and representative of the great German house of Rothbarth & Sons. Being entirely free from interests connected with hop culture, this opinion may well be taken as unbiased.

FIG. 127. INTERIOR OF HOP KILN.
Showing slat floor, with carpet rolled to end. Also shows upward sliding door (open) through which the dried hops are pushed into cars, shown in Fig. 112.

In the spring of 1897, an Oregon hop grower sold by contract to a merchant his growing crop of hops, which he guaranteed to be choice. Unfortunately the 1897 hop crop of Oregon was blighted and every sample showed traces of the blight, or "mold," as it is termed in this country. However, the particular lot of hops in question were as good as any that were grown in Oregon that season, and were manifestly therefore such hops as would have been selected or chosen by an expert who wanted the first quality of 1897 Oregon hops. Therefore they were choice hops, using the word choice as a noun, and constituted a good delivery on the part of the grower.

On the other hand, the merchant made a sale of choice Pacific coast hops to a brewer, without specifying any particular lot of hops, and inasmuch as the hops which he obtained from the farmer were speckled by blight, they were not of such a character as could be correctly designated by the use of the adjective choice, and hence he was unable to deliver them on that particular contract.

Summing up the whole matter, it is safe to conclude thus: To arrive at the grade of a given hop, it is best to say that that hop which in an open market should fetch the very highest price, would grade "fancy," while those fetching the lowest average price, would grade "common," and the difference between the highest price and the lowest price would be equally divided to mark the grades of choice, prime and medium. There is no other feasible way of describing or ascertaining the difference in market value between hops of different grades, as the true and only test (from the standpoint of grower and dealer) is the market value, which is based upon the consideration of all the desirable and undesirable features of a given hop.

SAMPLING HOPS PREPARATORY TO SELLING

While the wholesale sampling of hops by dealers has become a great nuisance to hop growers, they do not know just how or where to draw the line, as by refusing a dealer samples, they may miss the sale of their crop at a good figure. Therefore, they are entirely at the mercy of all the sample fiends who swarm over the country about the end of hop harvest, cutting and mutilating the growers' nicely sewed bales and scattering hops about. In giving a sample where the grower expects to make a sale, it is best to require the one who takes the sample to draw two from the same bale, the grower keeping one for the purpose of reference.

In case the dealer buys the hops and grades them from the sample in his possession, the grower can refer to his sample if there is any dispute. Hops in the bale change in storage, and a sample taken from a bale will sometimes look very different when compared with the hops in the bale some months later. In delivering, the grower should see that the buyer examines and grades all the bales purchased before marking any of them. This keeps the bales clean and free from marks, provided the parties cannot agree upon the proper grading of the crop, and does away with disputes that may result in lawsuits.

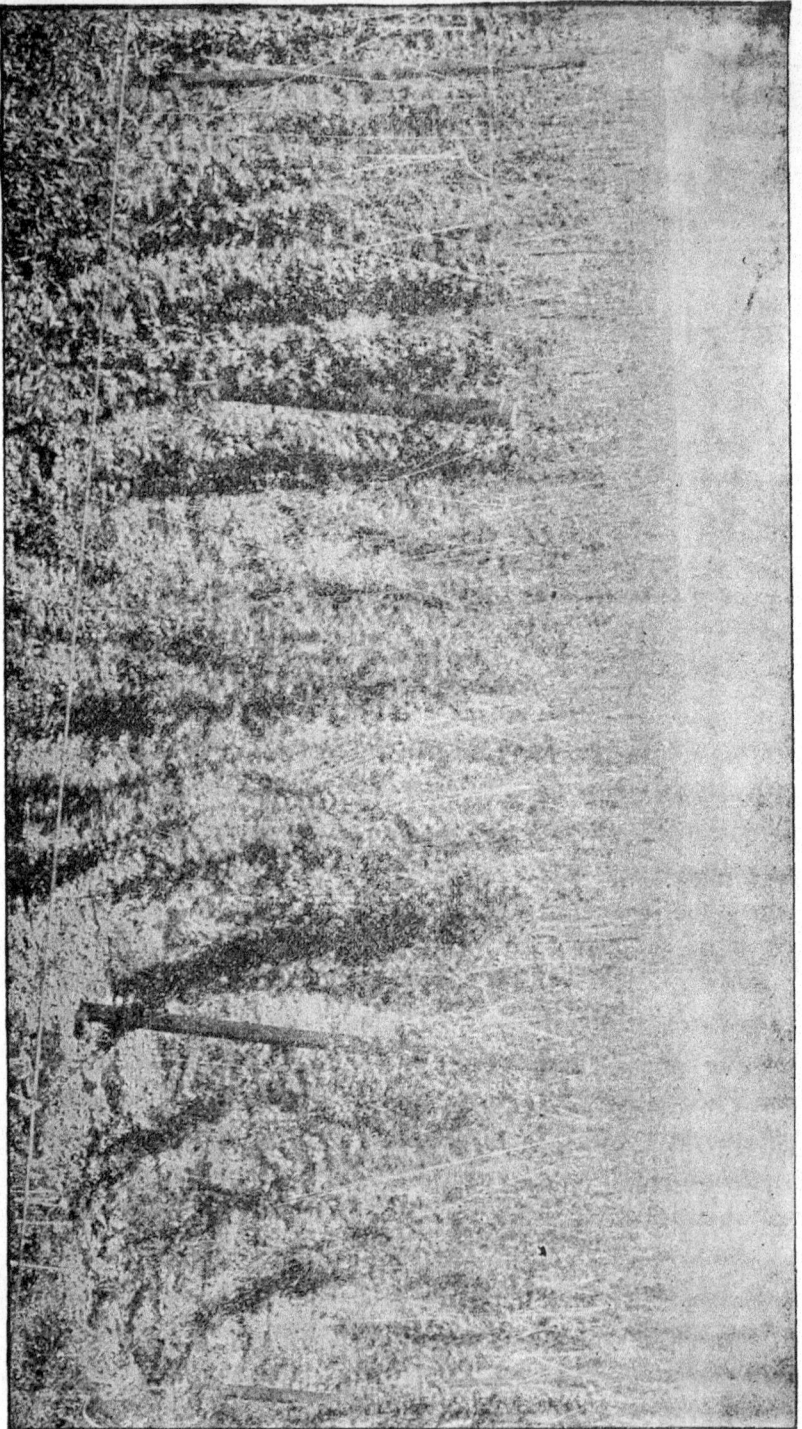

FIG. 128. BOHEMIAN HOPS FROM IMPORTED ROOTS.—See Page 216.

From photograph of one of Horst Brothers' Hop Ranches.

It should be universally recognized that sampling is "a delicate operation, requiring great care and nicety." A clumsy sampler will seriously injure the appearance of a lot: a clever man will give a nice "face" to the sample and leave the bale in nice shape. Whitehead describes a proper sample as follows:

"In a perfect sample the cones, as seen on the face, should be whole, with the strings or stalks completely free from moisture, and the lupulin or 'gold dust' adhering to the bracts. A very few leaves should be seen, and the cones should be single and not in bunches, and of a pale-gold color. An aromatic odor should pervade it, without the slightest trace of the sweet, 'gingerbready' smell, like heated clover hay, indicative of too much fire. Upon rubbing down some of the sample in the hand, there should be no fibrous residue, but the whole should chaff finely, leaving a yellowish, resinous deposit on the fingers. A well-managed and properly desiccated sample is most elastic, and can be compressed by the hand into a small compass, rebounding to its original size when the compression is removed. This is a valuable indication of judicious drying."

In a previous chapter we cautioned growers against mixing poor hops, if they have any, with their better ones. If a grower has more than one grade he should keep them in separate bins in the cooling room, and likewise bale them separately, as well as keep them separate afterwards, so that when a buyer samples them he will find the different qualities and offer on them accordingly. To mix a poor grade with a better quality will usually reduce the selling price of the lot to the figure at which the poorer grade alone would sell. Hence, keep the moldy, light-colored or otherwise inferior quality separate from the better grades. But with a nice crop of quite uniform appearance, it is generally desirable to carefully mix the entire growth so thoroughly that every bale will be of the same uniform grade and quality. It is often a question, also, whether it is not best to mix two or more varieties (we do not here mean grades of quality), for very often a careful mixture of hops from red vines with "clusters" is a decided benefit in flavor, as well as in appearance and attractiveness to the dealer or brewer. So, too, of the Humphreys, that are usually very bright, but which if mixed with the main crop help the average color of the whole without losing any market value on the Humphreys.

All hops are sold by sample or grade and subject to inspection, and it is customary for the buyer to take all the hops that are equal to or better than the sample or grade; while, at the same time, he has the right to reject any and all bales that do not come up to the selling sample or grade. There is no averaging of hops, and the buyer can be compelled to take only those hops that run equal to sample, and even if he secures some that are better, he cannot be compelled to take those that are worse. It can be readily seen that if a buyer is given a sample of the best portion of a crop, trouble will result when it comes to inspection. On the other hand, should the grower submit his worst sample, he will not receive what he is justly entitled to for his entire crop, as no buyer will object to hops that are better than the purchased sample.

Therefore, if a grower will keep his different qualities separate and submit samples of each quality to the buyer, and sell accordingly, there will be no misrepresentation and no chance for dispute. The ordinary grower will usually do well to have as few grades as possible, at least two—the main crop and late picked.

It is customary when a trade is made between buyer and seller, for the former to take his purchasing sample or samples and lay on a board or table beside the lot he has bought. He then proceeds to examine each bale with a tryer, which is an instrument about 16 inches long and three-eighths of an inch in diameter, with a crosspiece at one end, about four inches long, for a handle. The other end comes to a sharp point and has a hook similar to a fishing hook. This tryer is thrust into the bale, the cloth of which has been cut to a length of about two inches, and the tryer is then given half a twist and withdrawn. The hook brings out a good-sized handful of hops, which the buyer proceeds to examine and compare with the purchase sample. Each handful of hops is laid on a board or table and compared. By this means one can get a correct idea of the contents of each bale, especially to determine its condition as to curing. It is sometimes necessary to examine a bale in more than one place, to look out for false packing, of which we spoke in a previous chapter.

Hops that will crumble when compressed in the hand are over-dried and probably high-dried. Their brittleness indicates that the moisture, as well as the essential oil, has been dried out of them. On the other hand, a hop that is not dried enough, or what is termed in the trade "a slack-dried hop," will be clammy when compressed, and will not only feel moist, but will stick together in a ball or lump. The sample of a properly cured hop will be springy and full of life, and when a portion is taken in the hand and compressed, it will be found to be very soft and silky, and, like a silk handkerchief, will rebound when the pressure is released. When rubbed in the hand, it will emit a pungent, aromatic odor. Springiness in the sample is necessary, for it indicates just the right condition. A bale a little boardy may keep till warm July or August weather, and then heat.

An expert can get a very good idea of the quality of a hop by the mere feeling of it, and when upon compressing a handful from a sample he finds that it is elastic and springy and rebounds without crumbling, it is one of the numerous indications to him that the hop has been well handled in the curing and baling. On the other hand, if it feels quite moist and clammy, it is an indication of slackness, and if there is considerable moisture, he will know that the hops will soon heat. Everything else being the same, the springy lot will have a fine flavor and the boardy lot lack in aroma.

FIG. 129. WASHINGTON HOP KILNS IN KING COUNTY, NEAR AUBURN.

The flavor of a hop is another important feature in determining its value and quality. A flavor that savors of burned bread, or, as Whitehead terms it, "gingerbready," is an indication of too much heat or fire, while a sour or musty flavor is an indication of a slackness, a defect that sometimes proves quite costly, as it quite often happens that slack hops commence to heat and turn black after they have been put in the cars, although they may have stayed in the warehouse for some time without heating. A slack hop will not stand shipment for any distance, especially for export. For that reason it is essential that the hops be properly cured, while at the same time, they should not be dried too much, as that destroys the active brewing principles.

MARKETING THE CROP

Even when the hop grower's crop is safely baled, in accordance with the most rigid rules as to quality, his troubles are by no means ended. When to sell is the next problem, to which no definite answer can be given.

The course of prices in the Nuremberg hop market for upward of fifty years, confirmed by the range of prices at New York city, warrants the conclusion that, as a general thing, prices average somewhat higher during September, October and November than during the next quarter, while the lowest prices usually occur in spring and summer. Yet, there are exceptions to this rule. In the fall of '97 hops opened at low prices, and by midwinter had quite doubled in price. The extreme fluctuations in prices in different years, together with the rapid and violent changes from month to month, are shown by the tables of monthly quotations in the appendix. These tables will amply repay the most careful study. The grower who will not study them for himself would not be benefited by our analysis of the price tables.

Certain it is, that the export demand has an important influence on values. If the foreign market is short of old hops, and the new crop abroad is inferior, either in quantity or quality, the active export demand is a brilliant factor in the domestic market. The great bulk of American exports is from New York, and the export movement from that port is therefore of vital and constant interest. The appendix tables throw a flood of light on the movement and its relation to prices. Comparing these data for the five years 1889-'93 inclusive, with the three years more recently, '94-6 inclusive, an interesting exhibit is obtained:

HOP EXPORTS IN BALES FROM NEW YORK CITY, AND PRICES
(in cents per pound.)

Totals for	Five years, 1889-'93.			Three years, 1894-'96.		
	Exports.	% of total.	Av. price. Cents.	Exports.	% of total.	Av. price. Cents.
Sept. Oct. Nov.	101,838	41	24.9	67,996	32	10.7
Dec. Jan. Feb.	96,595	39	24.6	100,988	48	11.4
Mar. Apr. May	22,742	9	23.6	33,704	16	9.3
Jun. July. Aug.	27,895	11	22.1	9,455	4	7.5

The receipts of domestic hops at New York city by rail are reported in *American Agriculturist* from week to week, together with the exports. If it appears that the exports are taking a large proportion of the receipts, this is usually a healthy sign for the immediate future of values. The moment shipments abroad fall off and domestic supplies accumulate at New York, dealers make this a pretext for hammering down prices. The appendix exhibit of receipts at New York city should be carefully studied, in relation to both the total crop at home and abroad, and our exports and imports. Then, by comparing the weekly and monthly statistics as the season advances, one may form some judgment of the movement of the crop.

The uncertainty as to prices makes hops a very fascinating crop to speculate in, and growers as well as dealers often cannot resist the temptation to speculate in their community. When prices advance, it is impossible to foretell how high they will go, and growers feel justified in holding, but when prices begin to decline, no one is anxious to buy, as there is no telling how low they will go. Therefore, the hop grower frequently sells too soon or holds on too long, and rarely indeed does he realize the top market price.

The author has made a close study of this matter for some years, and has collected the actual experience in selling of about 100 planters, including men of all shades of ability in raising and selling hops. It appears that those who have sold early for cash at the market price, say between October 1 and December 1, have averaged better returns than their speculating neighbors.

The formers are still raising hops and are tolerably satisfied with the business; the latter have quit in disgust, many of them, if they have not failed outright. Growers who make every effort to put up a nice crop in good shape are mostly disposed to sell within three or four months of harvest at the going price. Their business is to make a nice hop; they are willing to let the men do the speculating whose business it is to speculate.

Aside from the matter of when to sell, about which opinions must always differ, there is an opportunity for growers to co-operate and greatly improve the method of selling. Yet, right here the division of sentiment as to the time of selling enters as a serious obstacle. The hop, however, is a commodity that can be readily sold by sample. If now the growers in each state or each section would unite in a hop exchange, through which their samples might be properly classified and guaranteed, this would be a great step in advance. These samples could be brought together at the office or headquarters of the exchange, and would thus attract the largest number of buyers. The sales could be made either privately or by the auction system, at stated hours or dates. By this method, the growers would bring together the largest number of buyers, thus creating a competition that would result in the best possible prices. This is quite different from the present method, by which the average grower too often feels obliged to accept whatever offer is made by the buyer who happens to come along when the producer feels like selling.

Apart from the division of sentiment over the time at which to sell, there should be no great obstacle against forming and operating such hop growers' exchanges. With no crop can this method of co-operative selling be handled to better advantage. The grower can fix his own price, and instruct the exchange not to let his crop go for less, and can modify his views either way from day to day as he deems fit. The exchange would be the headquarters for all information about crops and markets. Since the exchange would have to guarantee that the quality of the bale would be up to sample, it would have to enforce the most careful packing and a rigid inspection. Yet, this is no more than the buyer now expects, and by the exchange system the grower would suffer far less from these exactions than at present. Each crop would be sold on its merits, or, if desired, the exchange could arrange for the mixing of crops so as to furnish large lots of an even quality. With good business management and proper support on the part of hop growers, such an exchange could not fail to be of great value to both buyers and dealers. While new to the hop trade, it is not a new method of selling produce, but has been long used and not found wanting. We believe it is only a question of time before the exigencies of the business and the good sense of hop planters will lead them to unite in efforts of this kind.

Permanent improvement in prices, however, will depend more on limiting the production than through any other agency. How to accomplish this is an extremely difficult problem. Since spraying for lice and mold has come into vogue, it is not likely that such absolute failures will again occur as have occasionally characterized the past. Without concert of action, every naturally short crop, with its consequent high prices, will be followed by an increased area throughout the hop-growing world, to be followed by another period of low prices. The hop planter is therefore the architect of his own fortunes and can reckon on good prices only when the acreage is kept down to a minimum.

FIG. 130. HOP KILN, PLEASANTON.

But the trouble is that the average planter expects the other fellow will go out of the business and that, therefore, he can extend his own acreage. Co-operation to reduce the acreage has failed heretofore, and the individual action just noted too often leads to fresh overproduction. The improved methods of holding hops in cold storage enable brewers to lay in large stocks during seasons of plenty and low prices, and this militates against improvement in prices more than was formerly the case. We find this quite as true in Europe and in England as in the United States, except that in those countries the area does not expand as rapidly as in the United States, particularly on the Pacific coast, where a good crop is obtained the first year from planting.

STORING HOPS FOR LONG KEEPING

Several methods have been recommended and used to protect hops as much as possible against the action of atmospheric air. Pressing them meets with steadily increasing favor in England, and is generally used in America, instead of treading them into bags, as is customary in Germany. This pressing is of decided advantage, but the hops must be well dried before they are pressed. It has been recommended to press the hops into pitched barrels instead of bales, and to store them in ice cellars (Scharr). Bing, of Nuremberg, presses them into square bales by hydraulic presses; the bales are then put into well-soldered tin boxes, and then are placed in well-pitched wooden boxes. It has been further proposed to press the hops into tin boxes, to close them hermetically, and to store them in a cold cellar (Neubecker).

According to Brainard's method of preserving them, they are well dried and packed in bags, and brought into a store room, which can be kept dark, dry and cool, and can be hermetically closed. For this purpose, the store room has double walls, and is provided with ice on the upper floor, in the same manner as Brainard's store cellar and fermenting cellar. The efficacy of cold storage is seriously questioned, however. The store room should be closed as much as possible against the air, and should be dry and cool. It should not be located directly under the roof, where damp air can easily enter, and a simple partition of boards is not sufficient for this purpose. The best plan is to build the store room with bricks or double-frame sides, between which is placed an isolating layer.

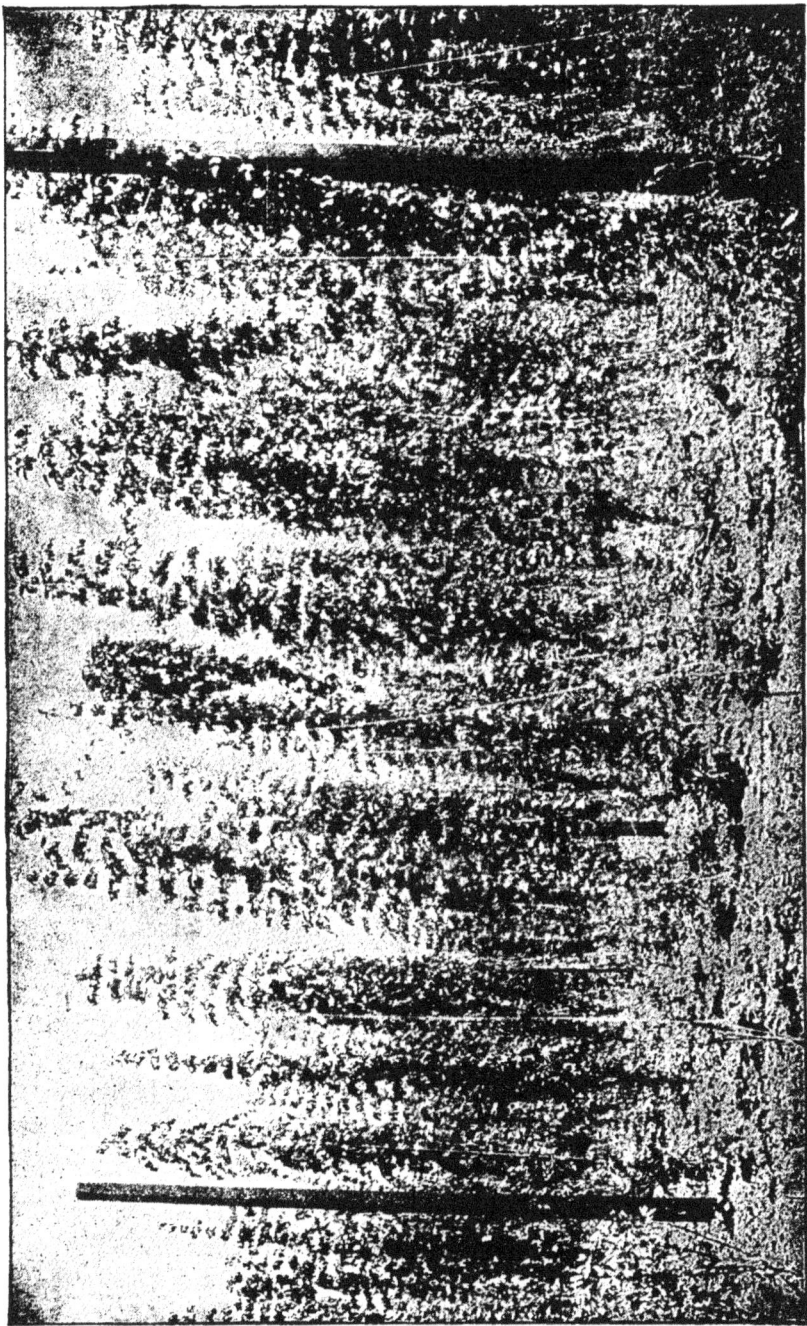

FIG. 131. BAVARIAN HOPS GROWN AT HORSTVILLE, CAL., FROM IMPORTED ROOTS.

East Kent hops, when four months old, contained 12.3 per cent. of hard resins and 29 per cent. of soft, but after having been stored a year in a brewery under ordinary conditions, the soft resins had fallen to 8.3 per cent. and the hard rose to 7.3 per cent. The chemist's theory that the brewing value of hops is gauged only by the soft resins, which is apparently borne out by practical experience in brewing, indicates a loss of one-third in the actual brewing strength of these hops during the year. If as new hops they were worth 24 cents per pound, then, as old hops, 16 cents would be a full price for them. This depreciation explains the importance of proper care of old hops. Hermetically sealed up in galvanized iron cases, hops have been kept for two years or more without appreciable loss in brewing power.

EXTRACTING THE LUPULIN

And bottling it has long been successfully practiced. One concern in central New York has made the fifth addition to its hop extract works, and has worked up more than 15,000,000 pounds of hops since its inception 25 years ago.

The process of preserving consists simply in removing the sacking from the bales, breaking the hops apart, putting them into large tanks, closing up the tanks, pumping in an easily evaporated solvent, which makes a solution of the lupulin, drawing off this solution into evaporators, where the solvent is evaporated and the pure extract of the hops left behind in the evaporator, whence it is drawn off and soldered up in cans, thus made air-tight, so that the extract will keep fresh for a good many years. This operation, waste, car freight, etc., cost five cents per pound of hops handled; 12 pounds of good hops yield one pound of extract, which is equal in the brewery to 12 pounds of hops. Brewers patronize this factory quite largely, especially when hops are dear.

A different method for accomplishing a similar purpose has been perfected at the municipal brewing school in Ghent, Belgium, and is now being used by a syndicate of Belgian brewers. They establish works near the plantations, for convenience of delivery, and buy hops only after analysis, paying according to the amount of lupulin they contain, quite regardless of color or odor. The hops are further dried at 95 degrees F. The dried cones are then operated upon by a machine, which, by the action of brushes, sieves and fans, breaks them up into their separate petals, and mechanically separates from them the lupulinic powder. The golden flour is collected and put aside. The leaves are then passed through a series of several washing cylinders, being treated therein with water at a specified heat, until they are completely macerated, and the tannic acid and other soluble matters contained in them are entirely removed. The undissolved proportion remaining after this process is very small; therefore the waste is reduced to a minimum. The fluid passing from the cylinders is then evaporated in vacuo at the temperature of about 95 degrees F., until it attains the consistency of syrup. To this are then added the grains of lupulin, forming together a kind of greasy, brown paste, which is packed in hermetically sealed tin boxes, from which the natural air is afterwards expelled, and replaced by carbolic gas.

In this manner it is claimed that all the useful preservative quality, as well as the flavor originally possessed by the hops is preserved without any deterioration, and that it can be sold to the brewer for his storage for an indefinite period. Other advantages shown to brewers are:—

Reduction in space required for storage, as the bulk of the hops is reduced by two-thirds, 100 pounds of natural hops producing 30 pounds of extract; uniformity of quality, whereby beer can always be produced with the same flavor, and containing the same preservative elements under any varying condition of climate; economy in using extract in comparison with natural hops. It is stated that the extract is completely soluble in cold as well as hot water, and that there is, therefore, nothing in it to affect the color or the clearness of the beer.

Putting on the Brimstone

FIG. 132. SCENE IN AN ENGLISH OAST HOUSE.

FIG. 133. A TRAINLOAD OF HOPS LEAVING HORSTVILLE. CAL.

CHAPTER XV.
CONCENTRATION IN HOP GROWING

In some respects the growing of hops on the Pacific coast is undergoing the same evolution that is witnessed in other industries. For lack of capital, low prices for the product, inexperience or lack of proper attention to the crop, many growers have been forced out of the industry, and in not a few cases have been obliged to sacrifice their plantations. Their farms thus become consolidated into large holdings owned and operated by men of ample means and knowledge, who conduct hop growing and marketing on a large scale, by scientific methods and on strictly business principles.

One of the most prominent instances of this tendency is afforded by the Horst Brothers. They have under cultivation a tract of 700 acres (see Fig. 6, Page 23) on their home ranch at Horstville, on the Bear river, Yuba county, Cal., and this one tract produces annually over 5000 bales, equal to a million pounds, of hops per year. They also own and operate other large tracts in hops on the Russian river in California, on the Willamette river in Oregon, and on the Fraser river in British Columbia.

The Messrs. Horst grow their hops against contracts that they have with brewers throughout the world, and that are made for a long term of years in advance, thus relieving themselves of the industry's speculative features and giving them control of a good share of the hop business. This plan has advantages for the breweries as well, as they are assured of a supply of hops of satisfactory quality at a fair margin over the cost of production, which is usually below what they would otherwise have to pay.

The ranch at Horstville is the basis for their extensive operations. The entire 700 acres devoted thereon to hops are set in the improved wire trellis on 20-foot poles that are set two feet in the ground, thus leaving the horizontal trellis wires 18 feet overhead, from which two strings run down to each hill.

FIG. 134. PICKING GOLDEN CLUSTERS

A yard is devoted to experimental purposes, where the different varieties from all parts of the world are tested and seedlings are originated, some of which bid fair to combine all the qualities most desired. The illustrations in connection with this chapter show a number of varieties most popular with brewers that are grown on a large scale on this ranch.

One remarkable feature about this establishment at Horstville is that lice and mold have never been known since hop culture was inaugurated there, forty years ago, and this fact, combined with a soil and climate peculiarly adapted to hop culture, makes this locality one of the most certain and most prolific sections for hops in the world. The ranch is on the banks of the Bear river, and in a dry season the river can be used for irrigating, thus making the crop certain regardless of rain. In the year 1898, when California suffered from such a drouth as had never been known, this hop ranch produced the same quality and quantity per acre as usual.

By thus concentrating, under one management, many plantations, the entire culture, harvest, curing and sale of the crop is in the hands of experts. Nothing is left to guesswork; slipshod methods are not tolerated. Every detail is conducted on businesslike and scientific principles and receives the benefit of the wide experience acquired by the owners of this ranch, not only in growing hops, but in disposing of them.

The latest improvement devised by Horst Brothers is their new kilns. These dry and cure the hops at the lowest possible temperature and are now being further improved with a system by means of which the hops are dried by currents of air driven through them by fan blowers. This air is not heated at all, and no artificial heat of any kind will be used to effect the drying or curing, thereby completely preserving the aroma, texture and lupulin,—qualities which are otherwise likely to be sacrificed, to a more or less extent, during the cure. These kilns and storage houses are built entirely of iron, bridge construction for the frame, corrugated iron for sides and roof, and the hop kiln floors are No. 4 steel wires, one and one-half inches apart, with No. 10 crosswires about six inches apart, thus giving practically all the surface to curing the hops, instead of only one-half, as by the ordinary wooden floor process. These iron kilns are considered so absolutely fireproof that no insurance is carried, and this style of construction is evidently to come into general use. All the other hop kilns on this ranch are now being remodeled to conform to the arrangement of the battery of six kilns above described, which are shown in the illustration on Page 193.

FIG. 135. EAST KENT GOLDINGS.
From photograph of one of Horst Brothers' Hop Ranches.

Another notable improvement used by Horst Brothers is a hydraulic compress for recompressing the ordinary 200-pound bale of hops into a package of one-half the usual size. This saves space in storage and in transportation, and brewers speak highly of the better keeping qualities of the hops thus compressed. The crop is here grown on such a large scale that it is shipped from the ranch by whole train-loads, to be distributed throughout the world.

FIG. 136. A PICKED YARD AT LEFT, UNPICKED TO THE RIGHT.
Pleasanton Hop Co., Cal. Turnstile of elevated tramway between kilns shown in the foreground.

CHAPTER XVI.
EXPENSES AND PROFITS OF HOP CULTURE

THE cost of growing hops varies widely, even between neighboring plantations, by reason of differences of methods and yields, and still more widely between different countries. Profits fluctuate even more seriously, depending upon both yield and prices. The yield of cured hops per acre ordinarily varies within the following limits as a fair average for all growers, but in extraordinary seasons may exceed them either way, while the best growers will often exceed the highest figure quoted:

YIELD OF CURED HOPS PER ACRE AND COST.
Pounds per acre.

	Highest.	Lowest.	Averages	Cost per lb.
Germany	800	400	500	20 to 30c
England	1,000	500	900	12 to 20c
New York	1,500	400	800	8 to 20c
Pacific coast	2,000	600	1,200	6 to 15c

GERMANY—Owing to the peculiar methods in Germany, previously noted, it is quite useless to attempt any statement of receipts and expenses of the peasant hop grower.

For ENGLAND, however, Whitehead, in 1893, revised all previous estimates, and Mr. E. H. Elvy, editor of the *Kentish Observer,* the leading hop journal in England, has carefully corrected the returns up to 1899 for this work, as follows:

The land on which hops are grown in Kent is worth about $200 an acre, and interest is reckoned at 5 per cent. To start a new plantation will cost from $100 to $125 per acre, including preparation of the soil, fertilizing, sets and planting, cultivating, rent, taxes, etc. Plants cost $1 to $2.50 per 100, usually $1.25. Poles or trellis cost $50 to $100 per acre. Thus the cost to get ready a new hop yard in England will vary from $150 to $275 per acre. The kiln for

20 acres costs about $2500, or, say, $100 an acre. After this investment, the following table affords a fair range of the yearly expense over a large part of the English acreage, being larger or smaller according to circumstances.

ENGLAND—AVERAGE YEARLY EXPENSE PER ACRE OF HOPS.[2]

	Highest.	Lowest.	Average.
Manure	$40.00	$20.00	$30.00
Digging	6.00	3.00	4.50
Dressing or cutting	2.00	1.00	1.50
Poling, tying, training, lewing	11.00	7.00	9.50
Cultivating and hoeing	16.00	10.00	12.50
Stacking, stripping, cleaning up yards	4.00	2.50	3.00
Annual renewal of poles or trellis	10.00	7.00	8.50
Picking, curing, packing, sampling, etc.	50.00	30.00	42.50
Rent, taxes, repairs, interest, etc.	32.00	25.00	27.50
Sulphuring to prevent mildew	5.00	2.00	3.50
Spraying again lice, etc.	12.00	7.50	9.00
Total	**$188.00**	**$115.00**	**$152.00**
Lbs. per acre under favorable conditions	1,000	700	900
Cost per pound, say	19c	17c	16c

Thus we get an annual charge per acre of $188 as one extreme, down to $115 as the lowest, or an average of $152 for yards kept up in good condition; against about $112 estimated by Marshall just a century earlier, about $120 by Mainwaring's figures in 1855, and Worcester planters' estimate in 1890 of $145 to $170 per acre for a good crop.

A—Annual cost per acre of hop culture in an East Kent yard, three pole system and simplest methods still largely followed.

Stripping vines and stacking poles	$1.25
Annual renewal of poles	25.00
Stable manure, also carting and spreading it	15.00
Digging $5, cutting $1.25, poling $3, tying $3	12.25
Digging about hills $1.25, shimming and harrowing $5	6.25
Picking $25, digging $10, pockets $6.25	41.25
Rent, rates, tithes $20, sundries $6.25	26.25
Total cost for average yield of 7cwt. per acre	**$127.25**
Average cost of hops per pound	16c

---

[2] The £ sterling is figured at $5, the shilling at 25 cents, the penny at two cents.

B—Annual cost per acre of hop culture in Mr. J. D. Maxted's yard, East Kent, on the Butcher wire trellis and highest culture by methods employed by the most enterprising planters in England.

Manures—12 loads of dung $15, 10 cwt. of artificial fertilizer $10, 5 cwt. gypsum $1.43	$26.43
Team labor—Carting out dung $4, plowing $2, twice 3-horse shimming $3, thrice 2-horse ditto $2.62, twice 1-horse harrow 50c. cartage on fertilizer 28c	12.30
Manual labor—Spreading dung 37c, digging slips three times $7.50, cutting $1.25, stringing $2.50, cutting off old vines $1, training $7.50, repairs to wire work and lews $1.50, digging round 50c	22.52
Sundries—Three cultivatings $11.38, three sulphurings $2.16, string for trellis $8.22, new poles and wire $2.40, new implements and repairs to old ones $3.60, blacksmith's bill $1.68, rent and rates $14.58, proportion of supervision $4.86, hire of oast $2.40	51.28
Total up to harvest time (about 6⅔ cts per lb.)	**$112.53**
Harvesting and marketing costs 30s per cwt., or within a fraction of 6½c per lb, including picking and curing and getting to market (coal, brimstone, pockets, binmen, tallymen, cartage to oasts and station, freight, insurance, commission and sampling), a total of	106.72
Aggregate cost of a yield of 15 cwt. per acre (or 1,680 lbs. at an average cost of a trifle over 13 cents per lb.)	$219.25

COST OF HOPS IN NEW YORK STATE, U.S.A.

NET PROFITS OF $150 PER ACRE—The late William Brooks of Cooperstown, New York, furnished a remarkable statement to the *New England Homestead* in 1885, in which he placed the cost of production at 10c per lb. He always gave his yard the best possible care and sold his hops at the market price when baled. He bought his farm of 100 acres in 1863 for $3000. It had five acres in hops, to which he added two more acres in 1866. From these seven acres, he received $38,180 for the 21 hop crops, 1863-'84, or an average of $2367 per year, equal to $339 per acre per year. But this was during a period when hops averaged higher than of late years. But the fact that the crop yielded an average of 1300 lbs. per acre all these years, or nearly double the product on neighboring yards, where cost per pound was as much or more, shows what can be done. His net profits must have averaged for the 21 years over $150 per acre. Mr. Brooks furnished details of his last 10 crops as follows:

	Bales.	Pounds.	Price per pound.	Tot'l rec'ts.
1875	48	9,910	13c	$1,288
1876	43	8,869	34c	3,113
1877	62	12,006	11c	1,309
1878	46	8,693	14c	1,141
1879	49	8,531	30-35c	3,512
1880	43	8,221	16c	1,315
1881	52	9,663	25c	2,417
1882	36	6,402	70c	4,481
1883	54	9,636	31c	2,876
1884	53	9,590	24-25c	2,223
Total	**486**	**91,521**	**av 25.8c**	**$23,675**
Av. per year, 7 acres		9,152	25.8c	2,367
Av. per year, per acre		1,307	25.8c	339

NEW YORK STATE—COST OF GROWING HOPS

OTSEGO COUNTY, N. Y.—W. H. G.'s 10-acre hop field cost $75 an acre, hills 8x8 ft, or 675 per acre; cedar poles at 12c, delivered, cost $810 for the yard. The kiln and store room is 50x24, and cost, including furnace, press, and other fixtures, $1600. The sacks for green hops, boxes for picking, etc., cost $40. The cultivators, hillers, grub hooks, bars for pole-setting, etc., cost $50. The pickers boarded themselves and at 40c per box were paid $206 for the 515 boxes; five box-tenders at $1 each for 15 days, $75; expense of collecting and carrying pickers, $15; man at kiln 15 nights, at $1.25 per night, $18.75; use of kiln-cloth $3 (it cost $45 for 900 lbs. at 5c), a total of $317.75, or 3.9c per lb. of cured hops. Insurance was $3500 on hop house for 30 days at 40c per $100, making $14, and $2000 for balance of year at $12. Work is charged for at its local market value. Total cost a trifle over 12c per lb., and as he sold for 13c, he made a slight profit over and above fair return for his labor and capital. The operating expenses were 8.5c per lb. sold, fixed charges 3.7c, or a total cost of 12.2c per lb. of hops. If $30 worth of the $88 spent for fertilizers is allowed to be in the soil for the next crop, the net operating expense of this 1897 crop was $663.25, or $66.35 per acre. Adding depreciation and taxes, $259, the total cost of production is $922 for the ten acres, or, say, $92 per acre. Deduct this total cost from the receipts for the crop, and the balance of $126 represents the net returns on the $750 invested in the land, or 17 per cent. Or, if we figure the investment at $3250 (including land, $750, poles $810, building $1600, tools $90), the difference of $395 between operating expenses ($663) and receipts ($1058) represents the net earnings on the investment, and shows a net income on such investment of nearly 13 per cent. In the table the items are arranged in the order that the work was done.

	Ten acres.	One acre.
Setting poles at 20c per 100	$13.50	$1.35
Grubbing by hand	25.00	2.5
Two-horse cultivator, once both ways	12.00	1.2
Tying up shoots, four women at 75c per day	9.00	.90
Twine and labor putting on	40.00	4.00
Tying, trimming, training (women)	30.00	3.00
Cultivating again, both ways	10.00	1.00
Fertilizing (four tons hardwood ashes at $12, one-ton bone meal $40)	88.00	8.80
Handling, mixing and applying fertilizers about hills	7.00	.70
Twine, and putting on	40.00	4.00
Hilling, shovel-plow one way and hoeing	20.00	2.00
Last two cultivatings (lightly), tying up broken vines, etc.	15.00	1.50
Harvesting and curing (details above)	317.75	31.77
Brimstone $3, fuel $8, insurance $26	37.00	3.70
Baling at 20c per bale	9.00	.90
Stacking poles, covering hills for winter	15.00	1.50
Delivering crop at station	5.00	.50
Operating expenses for crop	**$693.25**	**$69.32**
Fixed charges (interest on land at 6 per cent., $42; depreciation of poles at 10 per cent., $81; depreciation on kilns, tools, boxes, etc., $169; taxes, $9)	301.00	30.10
Total cost of crop	**$994.25**	**$99.42**
8140 lbs. hops sold and netted	**1,058.20**	**105.82**
Net balance	**$53.95**	**$5.40**

MADISON COUNTY N.Y.—L. W. Griswold gives his estimate of cost of raising one acre of hops in the table below. Dividing the total cost by 1000 lbs., which is certainly a large average yield per acre, it gives the cost of the first crop as a little over 25c per lb., exclusive of the cost of building kiln and storehouse. Deducting the price of poles, roots and tools, preparation and planting of yard, and adding $5 for the breakage of poles and wear of tools, we find the cost for the following year to be $76.85, or a trifle over 7½c per lb. When, however, we add $500 for budding a kiln and storehouse, to the other expenses, and depreciation, interest, etc., on same the actual cost is far above 7½c.

Preparing ground for planting	$3.00
Sets for planting 750 hills	7.50
1,500 hop poles, 10c each	150.00
Tools, including two hop boxes	12.00
Setting poles, 20c per 100	3.00
Grubbing, one day's work	2.00
Plowing and cultivating twice each	6.00
Hoeing twice, two days' work	4.00
Tying three times	3.00
Picking 77 boxes, 30c per box	23.10
Boarding and lodging pickers	12.00
Drying hops	5.00
Baling 10 bales, 50 yds. Sacking and labor	7.50
Interest on land, valued at $150 per acre	7.50
Insurance $2, taxes 75c, fertilizers $6	8.75
Total cost	**$254.35**

ANOTHER OTSEGO STATEMENT (by James Ferris)—The largest grower in the county failed, though his hops sold at an average of 20c; another whose real estate was free of mortgage in 1893 failed in '97. Hemlock poles 18 to 25 ft. long cost 11½c, delivered ready for setting, 851 per acre, one to each hill 7x7 ft., or $97.86; with proper care, they last 15 years, annual loss, $6.52, interest at 6 per cent., $5.87, total yearly cost of poles $12.39. It is easy to determine cost of setting a yard and first year's cultivation. Potatoes or corn, potatoes preferably, are planted with hops the first year, occupying three-quarters of the ground. As 120 bu. of potatoes to an acre is an average crop, the hops would displace just 30 bu. of potatoes per acre, which, at an average price of 40c per bu., would have been worth $12. But the seed for hops usually costs more, and they receive better care than potatoes. Such additional cost is about $2 per acre. The average period which a hop yard will last and be productively profitable is about six years. So that the average cost of planting yards to displace those running out would be $2.33 per acre per annum on all hop land harvested. During depressed times, only one shovelful of barnyard manure is placed in each hill in the autumn, but when prices are good, more is used—about eight two-horse loads are used per acre, worth $8, and it costs $2 to apply. When pickers are plenty, they can usually be hired to pick and board themselves for 40c per box at present (1898), but in this locality not half enough pickers can be hired to pick and board themselves. The grower is obliged to board them, and go some distance after them, making the average cost of picking about 45c per box, and as hops usually cure about 15 lbs. to the box, this would make the cost of picking 3c per lb. Drying can be hired done at ¾c, the grower finding brimstone and fuel, and this is as cheap as he can do it himself if due allowance is made for capital, depreciation and insurance of kiln. My figures make the crop of 700 lbs. per acre (which is about the average) cost 13¼c per lb., as follows:

Hop poles $12.39, renewing roots $2.33	$14.72
Manure, and its application	10.00
Clearing up in fall, stacking poles, etc.	2.50
Setting poles $1.50, grubbing $1	2.50
Plowing four furrows per row, each way from hill	2.50
Cultivating twice in row both ways	1.00
Tying twice $2, twine $4, putting on $1	7.00
Training and hoeing	4.00
Winding on twine, tying with ladder	2.00
Plowing to hill $2.50, hilling $1.50, cultivating $1	5.00
Picking 3c per lb. cured hops, box tending and yard boss 1c, custom drying ¾c, brimstone, fuel, baling, marketing ½c, total per 700 lbs.	26.75
Baling cloth $1.40, insuring crop 42c	1.82
Rent of land at least	3.00
Total for 700 lbs. hops, one acre	**$82.79**

ST. LAWRENCE COUNTY, N.Y. (S. Hemingway)—Small items might be added to make the total below an even $110 per acre, or 20c per lb. for 500 lbs. per acre, about 11c for 1000 lbs., or (allowing for heavier manuring and increased cost of harvesting and curing) about 7c per 2000 lbs. per acre. The kiln, 20x30 ft, cost about $300, on which interest, $18, depreciation, $20, and insurance, $10, cost $48 per year, one-fifth of which is charged against one acre. I use two poles to each of 680 hills per acre, or 1360 poles per acre, costing $68, and allow 10 per cent. for depreciation.

One-fourth day uncovering hills 35c, two days setting poles $2.50, one day plowing $2.00	$4.85
One day hoeing $1.25, one day tying $1.25, one day trimming $1.25	3.75
680 lbs. fertilizer	10.40
Three days' plowing, 2d, 3d and 4th times	6.00
Three days' hoeing $3.75, two days resetting poles $2.50	6.25
One-half day's attention weekly for 16 weeks	10.00
Picking 1,000 lbs. at 3c	30.00
Two nights' drying at $2	4.00
One-half day's baling, two men	2.50
Two days' stacking poles $2.50, one day's cutting and burning vines $1.25, one day's covering hills $1.25	5.00
Total operating expenses	**$82.75**
Fixed charges: Interest at 6 per cent. On land worth $50 an acre $3, depreciation on poles $10.88, use of kiln $9.60	23.48
Aggregate expense per acre	**$106.23**

FROM NORTHERN OHIO

Statement of Danner and Hatch of Richland County, Ohio.

The plant—buildings, tools, boxes, press, etc. for five acres cost $245, interest on which is $14.70. As it lasts 20 years, 5 per cent. is allowed for depreciation, or $12.25 per year, or $26.95 per annum for use of kiln, of which one-fifth is charged to the one acre. One yard lasts about six years on our gravelly clay loam, rolling and fairly drained; worth $50 per acre, tax 50c. Crop for four years (1893-'97) averaged 80 boxes, or 960 lbs. per acre; set 7x7 ft., or nearly 750 hills per acre. First year's expenses are $24 (of which one-sixth is charged up annually), and includes 10 loads of manure $5, fitting ground $3, roots $1, planting $3, cultivating and hoeing four times $6, five loads manure to cover hills $2.50, interest and taxes $3.50, poles (two to a hill) 1500 per acre, cost 2c each delivered, or $30, and being good for six years, cost $5 per year. The annual expenses in the second and subsequent years will average as follows:

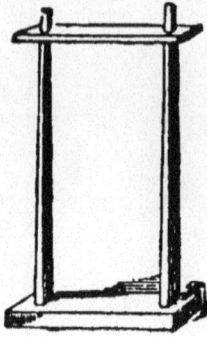

a

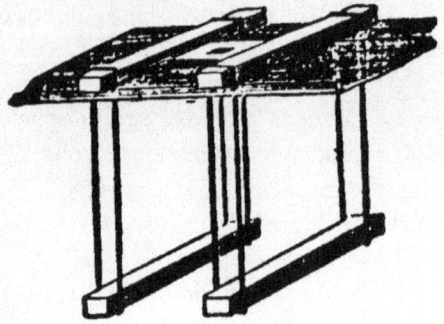

b

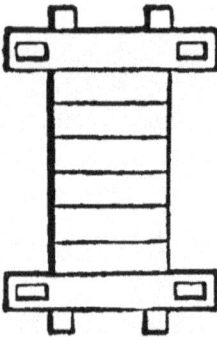

c

d

FIG. 137. HOMEMADE HOP PRESS.

a, Studs to hold press together; *b,* manner of supporting press; *c,* bottom of press; *d,* finished press. Four rods of 3¼ in. iron about 16 ft. long are bent, and ends welded together like large chain links. These are passed through the floor above the press, where they are supported by strong scantling 4x4. In the lower ends place timbers 4x6 as bed pieces, *b.* The bottom, *c,* is made of 2 in. planks 2 ft. long, with end strips 3½ ft. in length. The studs, *a,* at the sides, are of 2x5 in. stuff, mortised into the bottom and held together by a long mortise at the top. The sides are shown in *d.* The box may be 5 ft. long inside, 18 in. wide, and 6 ft. high.

Annual charge on first cost	$4.00
Grubbing, two days' work at $1.50	3.00
Poles $5, and poling 3½ days $5.75	10.75
Sharpening poles (40c per 100, good for 3 yrs, $6), one year	2.00
Plowing both ways, 1½ days at $2	3.00
Cultivating four times at 75c	3.00
Hoeing twice, four days, at $1.50	6.00
Tying up vines 1½ days at $1.25	1.87
Picking 80 boxes hops at 25c	20.00
Board 28 pickers 168 meals at 10c	16.80
Four box tenders, two days at $1	8.00
Board box tenders, 24 meals at 10c	2.40
Put hops on kiln, two kilns at $1	2.00
Man to dry, two kilns at $1.50	3.00
Wood 2½ cords, two kilns at $1.50	3.75
Brimstone, two kilns	.90
Baling five bales at 30c, delivering 50c	2.00
Sacking $1.25, stacking poles $1.50	2.75
Total operating expenses	**$95.22**
Fixed charges: Depreciation $5.93, rent $3, taxes, 50c	9.43
Aggregate (10.8c per lb. for 960 lbs.)	**$104.65**

COST OF HOPS ON THE PACIFIC COAST

CALIFORNIA—Daniel Flint says a hop kiln for 50 acres with all things complete will cost $3500 to $4000. High wire trellis costs $80 to $90 per acre, 2000 roots $20, and Japanese will contract to do for $10.25 per acre all the hand labor on the crop until it is ready to harvest. Picking, curing and baling costs him $2200 on 64 acres, and $1500 on 40 acres, an average of about $35 per acre.

OREGON, Washington Co.—E. C. Malloy submits a statement of a nine-acre hop yard started in 1893 on land worth $25 an acre, interest at 10 per cent, taxes 17 mills on the dollar. No manure is used, nor hoeing after the first season; kiln is 25x25 ft, 20 ft studding, that cost $200, two furnaces and pipes $54, warehouse $125, total on building $379. The harvesting equipment, picking and curing the first crop cost $51. Plowing, setting out, cultivating and poles for the first crop, produced the same year, $281, interest and taxes $29. This makes an even $1200 for cost of first crop, or $133 an acre for a yield averaging 1300 lbs., which would have to net 11c per lb. to pay all these expenses and leave the yard in good shape.

For the next crop it cost $1.25 per acre for cleaning up yard, $11.25; $12 per acre for setting poles, tying up hops and cultivating, $108; harvesting, curing, baling, etc. $472; interest, insurance, taxes, and depreciation on the whole outfit, $107. This made the second crop cost $698, or over $77 per acre, equal to 6c per lb. on 1300 lbs. per acre. "To further show the uncertainties of this business, especially in this region, I want to say that instead of getting 11 and 6c for those crops, I got 5⅞ and 4c, while many others consigned their hops on advances of 2 to 2½c per lb. and never got another cent."

OREGON: A Polk Co. Report—For the first plowing in spring, one man and two horses will plow four acres a day, at $2 per day, which is 50c per acre. One man and one horse will cultivate down five acres a day at $1.50 per day, or 30c per acre. Cross plowing will cost the same as first plowing, and cross leveling the same as first cultivating. With disk harrow, one man and two horses will go over five acres a day at a cost of 40c per acre; three additional cultivatings will cost the same each. Smoothing, or clod smashing, both ways, one man and one horse, five acres, or ten acres one way, will cost 30c per acre. It takes 12 lbs. of 10-ply cotton twine for an acre, at 12c per lb. Putting on twine, one man, five acres one way, costs 40c per acre twined both ways. Land is worth $75 an acre, interest 8 per cent., taxes $1.50 per acre, repairs and depreciation on tools $1.25. Three sprayings will require 18 lbs. quassia chips $1.08, 36 lbs. whale-oil soap $1.80, labor $2, repairs 37c, total $5.25. This gives us for one acre:

Cultivating as above	$3.60
Hauling 40c, and setting stakes $2	2.40
New stakes $1.50, twine and twining $1.84	3.34
Hoeing and sprouting hills	3.00
Training vines four times	6.00
Pruning surplus vines	2.00
Cleaning yard in fall	1.25
Interest and taxes	7.90
Spraying three times	5.25
Total (about 2⅓c per lb. for 1,500 lbs. of cured hops)	**$34.74**

Picking at 40c per box will cost about 3c per lb., and yard help (including delivering hops at kiln) ⅛c more. Drying hops, including wood and sulphur, 1c; pressing including hop cloth, five yards to the bale, at 8c, ⅜c; insurance, interest and repairs on plant, warehouse storage, etc. 1c, making a total of harvesting expenses of 5½c a pound. The crop will therefore cost the grower about 8c per lb. These figures are not the itemized expenses of any one grower in any particular year, but will cover the average cost for the last three years (1896-7-8) of those who own and work their own hop yards. "I do not think that any one man's itemized expenses for any one year is a safe basis from which to draw conclusions, as my hops have not cost me exactly the same any two years."

OREGON, Yamhill Co., J. W. F.—My hop garden contains 20.37 acres, valued at $125 per acre. The hills are eight feet apart each way, a total of 12,915 hills. The poles are fir and cost on the yard two cents apiece, or $258.30. The kiln is an octagonal building, 28 ft. each way in the clear, with storeroom combined 20x24 ft; with furnace, piping and press, it cost $910; 150 sacks $30, five measuring boxes of cedar, holding nine bushels each, $4.50. We use no fertilizer, as to cultivate the ground well is all that is necessary here, the soil being of a clayey nature, mixed with a very small amount of sand. Hops at this date (Feb. 14, '98) are nearly all sprouted or up, now and then one an inch long. My '97 crop was 16,187 lbs., or an average of 795 lbs. per acre, and cost a fraction over 7c per lb., as follows:

Cleaning up and burning vines	$14.00
Grubbing $35, setting poles $27, twine $16.90	78.90
Putting twine on poles $7, training and hoeing $158, plowing $58, harrowing $9	232.00
Rolling $11, reversible disk harrowing $22	33.00
Spraying	46.00
Picking 1,258 boxes at 40c	503.20
Yard man during picking	27.00
Two men to measure hops	39.00
Man and team to haul green hops to kiln	28.00
Two men at kiln 11 days at $2 each	44.00
440 yards of baling cloth	37.40
Baling 88 bales at 20c	17.60
Twine to sew up bales	2.70
Kiln cloth $5.10, 600lbs. sulphur $10, fuel $10	25.10
Oil for press and lights	2.00
Two men and teams to draw hops to station	5.50
Insurance and taxes	34.53
Total cost	**1,169.93**
Balance net profit	934.38
Total crop sold at 13c per lb.	**2,104.31**

WASHINGTON, KING COUNTY, 1897 CROP
(BY ALEXANDER ADAIR).

	Ten acres.	One acre.
Grubbing $50, setting poles $50	$100.00	$10.00
Tying up vines	25.00	2.50
Plowing and cultivating	100.00	10.00
Spraying, three men and horse 10 days	50.00	5.00
Quassia chips, whale-oil soap	36.00	3.60
Picking, $1 per box	456.00	45.60
Six men eight days at $2	96.00	9.60
Insurance on hops and kilns $2,000	55.00	5.50
Freight on hops to Seattle	45.45	4.54
Hop cloth and sulphur	41.00	4.10
Total	**$1,004.45**	**$100.44**
Harvested and sold, lbs.	11,700	1,170
Sold at 8c per lb	$936.00	$93.60
Loss on crop	68.45	6.84

RAISING THE CROP IN THE NORTHWEST

BRITISH COLUMBIA (Major R. M. Hornby)—Before going into hop culture, the novice should realize that it is one of the most uncertain of crops, that two good crops, three medium and two failures can be reckoned on every seven years, both as to yield and value. Only the best hops are now wanted. Such require the best land, outfit and methods. The cost of starting is large, and for an eight-acre yard (yielding 1800 lbs. of cured hops per acre in a favorable season) may thus be estimated.

Hop kiln, 24x24 ft., with stove complete	$1,250
Poles, 7 ft. apart, 820 per acre, at $30 per 1,000	177
1,640 hop sets per acre, at $3 per 1,000	40
Marketing out yard and planting sets at $5 per acre	40
100 hop boxes of 15 bu. capacity	100
Two double-acting spray pumps, with barrels and sleigh	50
Hop press	165
Total first cost	**$1,822**

The annual charges include interest and 15 per cent. depreciation on the above items of first cost, together aggregating $325 a year. Good hop land is worth at least $100 an acre, and interest and taxes may be added to the following figures. No insurance is included because the rate is too high, and proper care is cheaper than to pay insurance. Neither do we use manure on our rich lands, on the Pacific coast, which saves a large item that eastern and foreign hop growers have to pay, and their yield per acre under favorable conditions is not as large as ours. When yard forms part of farm, horses and implements are not charged to initial expenses, because they are part of the farm outfit. With this explanation we get the following:

ANNUAL EXPENSES ON EIGHT-ACRE YARD.

Depreciation and interest	$325
Setting up poles at $5 per acre	40
Credit the farm for one man, two horses, with use of implement for all horse cultivation and work	250
Tyings at $3, $1 and 50c per acre	36
Spraying once $4 (may be $12), say $7 per acre	56
Picking 1,800 lbs. per acre at $1 per box	504
Curing and drying	90
Baling by four men two days $12, floor and baling cloth, string $30	42
Cleaning up yard at $2.50, hauling to depot $1	27
Total annual expenses	**$1,370**
Profit if all goes well	730
Receipts for 1,800 lbs. per acre at 15c on eight acres	**$2,100**

Sometimes the yield is larger, more often less. The price is oftener less than more. The above makes an expense of about $171 per acre, or about 10c per lb. on a good, full crop. But the expense up to harvesting is the same, whether the yield is large or small, the quality good or bad. With the wide fluctuations in crop results, it is easy to see that cost per pound of hops in the bale may easily mean far above 10c per lb. and seldom below it. With market prices ranging from 5c to 25c, the speculative nature of the industry is apparent, as many have learned to their sorrow.

APPENDIX

STATISTICS OF THE HOP TRADE

RECEIPTS OF DOMESTIC HOPS AT NEW YORK (IN BALES.)

Crop of	1890.	1891.	1892.	1893.	1894.	1895.	1896.	1897.	Av.
September	8,374	5,572	2,926	9,305	5,108	3,216	2,778	3,388	5,092
October	24,809	20,200	14,376	25,399	26,466	22,086	16,836	15,074	20,656
November	23,411	27,386	19,882	31,669	32,339	36,015	34,712	21,190	28,325
December	6,430	24,242	23,302	24,141	30,088	22,028	13,930	30,626	21,848
January	4,778	15,775	13,819	10,595	21,236	17,495	7,297	25,408	14,550
February	4,114	8,954	5,162	7,883	12,100	15,257	6,565	8,125	8,145
March	5,636	5,597	8,361	6,792	11,340	13,215	2.911	5,802	7,456
April	2,697	5,842	6,316	5,418	7,051	4,644	2,998	3,844	4,826
May	2,789	2,239	6,583	5,704	8,749	3,179	2,039	1,569	4,106
June	3,789	964	10,119	5,423	3,282	3,432	2,222	2,195	3,928
July	3,320	1,885	9,628	4,629	5,361	2,515	1,721	1,500	3,819
August	2,989	1,767	8,664	4,335	3,049	1,348	1,440	1,952	3,193
Tot. rec'ts	93,136	120,423	129,138	141,293	166,241	144,430	95,449	120,673	126,348
Exported	22,804	54,619	64,205	74,623	83,749	76,506	51,892	87,165	64,445
Dom. use	70,332	65,804	64,933	66,670	82,492	67,924	43,557	33,508	61,903

EXPORTS OF HOPS FROM THE PORT OF NEW YORK (IN BALES)

Crop of	1890.	1891.	1892.	1893.	1894.	1895.	1896.	1897.	Av.
September	2,086	3,681	1,007	4,223	1,218	800	513	3,010	2,067
October	7,083	4,748	8,653	6,890	7,927	7,875	8,625	3,976	6,972
November	5,540	16,393	7,615	21,217	10,692	16,390	13,956	10,754	12,794
December	1,271	18,260	10,697	14,028	21,970	19,858	11,326	23,459	15,108
January	698	6,376	12,497	10,102	13,260	9,084	5,626	25,526	10,396
February	1,041	2,124	2,287	3,867	8,404	7,973	3,487	11,372	5,069
March	1,405	1,776	1,357	2,578	9,183	9,094	4,956	4,554	4,350
April	218	578	4,176	2,530	4,448	2,351	856	948	2,013
May	946	19	2,377	2,677	1,083	1,380	357	785	1,028
June	640	18	4,672	2,713	2,964	1,319	947	1,123	1,799
July	1,474	—	5,146	2,872	1,442	237	844	962	1,854
August	402	646	3,721	926	1,158	145	399	716	1,014
Tot. N.Y.	22,804	54,619	64,205	74,623	83,749	76,506	51,892	87,185	64,448
Tot. U.S.	49,000	70,000	64,000	97,000	97,000	93,000	63,000	95,000	78,000
% via N.Y.	47	78	100	77	87	82	82	91	85

AVERAGE MONTHLY PRICE PER POUND

In cents, of choice state hops at New York City. The periods are for the years inclusive.

	Sept	Oct	Nov	Dec	Jan	Feb	Mar	Apr	May	Jun	July	Aug
1874-1896	.252	.255	.267	.267	.258	.250	.235	.229	.226	.216	.21	.202
1874-1878	.254	.251	.252	.245	.236	.224	.201	.187	.186	.183	.178	.18
1879-1883	.346	.375	.435	.459	.442	.429	.416	.409	.396	.367	.350	.322
1884-1888	.250	.236	.225	.210	.192	.183	.175	.167	.164	170	.178	.181
1889-1893	.251	.249	.247	.241	.248	.248	.230	.235	.243	.234	.219	.209
1894-1896	.10	.103	.117	.123	.113	.108	.10	.091	.09	.08	.076	.07

FLUCTUATIONS IN PRICES OF CHOICE STATE HOPS.

AT NEW YORK CITY. HIGHEST (H) AND LOWEST (L) PRICES IN CENTS PER POUND.

Crop of	Sept. H	Sept. L	Oct. H	Oct. L	Nov. H	Nov. L	Dec. H	Dec. L	Jan. H	Jan. L	Feb. H	Feb. L	March H	March L	April H	April L	May H	May L	June H	June L	July H	July L	Aug. H	Aug. L
1874	39	39	44	39	45	45	45	45	48	47	47	47	44	40	40	39	40	39	40	37	37	37	37	37
1875	19	19	19	15	15	15	15	15	15	15	15	15	18	17	18	18	18	18	18	18	18	17	20	17
1876	40	40	40	37	37	37	30	30	30	26	26	20	20	16	17	16	18	17	16	16	13	13	13	12
1877	14	14	14	13	13	13	13	13	15	15	13	13	17	16	12	10	15	14	15	15	16	13	13	10
1878	15	15	15	15	15	15	15	15	13	13	13	13	13	12	9	8	8	8	8	8	8	8	8	8
1879	36	36	40	36	40	40	41	41	41	40	40	40	40	38	38	36	36	36	36	36	36	35	35	29
1880	29	29	28	25	24	24	24	23	23	23	23	23	23	23	24	24	26	23	23	23	22	22	29	22
1881	23	23	29	23	29	29	30	30	28	28	28	27	25	25	23	23	26	26	26	26	44	38	44	27
1882	58	58	80	60	1.10	80	1.13	1.02	1.02	1.02	1.00	94	94	92	96	92	96	80	78	52	52	38	50	44
1883	27	27	27	27	27	27	27	27	28	27	27	27	27	27	27	25	25	25	32	25	35	32	34	27
1884	26	26	26	22	22	22	28	28	28	28	28	27	27	27	27	25	25	25	25	25	32	32	32	26
1885	12	12	12	12	12	12	11	11	11	11	11	11	11	11	11	11	11	10	12	12	15	15	12	10
1886	30	30	30	28	30	28	30	29	29	26	26	24	24	22	23	23	21	20	25	20	25	19	23	20
1887	25	25	25	21	21	21	17	17	17	15	15	14	14	14	13	13	13	13	13	13	13	13	14	13
1888	32	32	32	28	28	27	23	23	23	22	22	22	22	22	22	22	24	23	24	23	23	22	21	16
1889	18	16	16	13	13	12	12	12	16	14	17	16	18	17	18	17	19	18	19	19	24	22	27	24
1890	48	48	48	45	48	42	38	38	38	38	38	33	32	28	32	28	32	32	32	29	28	28	28	24
1891	16	16	17	16	21	18	21	21	22	22	28	23	23	22	26	26	32	30	25	25	25	25	28	27
1892	23	22	26	22	24	24	23	23	25	23	25	23	25	25	22	22	22	22	23	22	23	23	23	23
1893	24	24	24	22	22	22	25	22	28	28	28	23	32	28	27	22	28	30	28	25	27	25	28	27
1894	10	10	10	10	13	11	12	12	22	22	22	21	22	22	22	21	17	17	19	15	15	15	20	18
1895	10	10	11	10	10	10	10	10	9	9	9	8	8	8	8	8	8	8	8	8	8	7	8	8
1896	10	10	11	10	15	11	14	14	13	11	13	12	12	10	10	10	9	8	8	8	8	7½	8	7½
1897	14	8	17	13	18	16	18	17	13	13	13	12	12	10	10	10	9	8	8	7	8	7	7	6
1898	13	11	21	17	21	19	21	19	20	18	19	19	19	17	19	16	17	15	16	15	15	12	13	12

EXPORT TRADE IN AMERICAN HOPS

Showing exports of domestic hops from United States each month for nine years, also imports and values. In thousands of pounds, last three figures (000's) omitted. The season 1895-6 was a period of low prices throughout. Total exports were 16,765,000 lbs., imports 2,772,000 lbs., average export price 8.8 cents, import price 21.6 cents. Prices at New York ranged at 7 to 10½ cents.

Crop year.	1897-8	1896-7	1894-5	1893-4	1892-3	1891-2	1890-1	1889-90
July	209	167	416	1,113	15	268	151	55
August	275	116	346	635	141	127	510	37
September	568	194	260	1,008	48	506	1,220	78
October	609	1,811	1,412	1,338	1,484	1,260	2,388	1,541
November	2,354	2,771	2,300	4,108	2,300	2,596	2,552	1,289
December	3,734	2,216	3,993	3,691	2,267	4,923	545	2,686
January	5,371	1,445	2,831	1,931	1,577	1,966	496	860
February	1,803	974	2,030	963	894	501	210	390
March	1,387	996	1,811	699	369	315	292	262
April	371	331	895	838	753	116	125	167
May	237	138	547	544	418	16	144	69
June	242	268	682	603	1,098	11	102	107
Total exp's, 12m	17,162	11,425	17,523	17,473	11,367	12,605	8,736	7,541
Total imp's, 12m	2,576	3,018	3,134	828	2,691	2,506	4,020	6,540
Re-exports, 12m	37	57	93	135	85	176	223	418
Net imports, 12m	2,539	2,961	3,041	693	2,605	2,330	3,797	6,121
Bond June 30	25	8	154	139	168	222	280	264
Total val. imp's.	$648	630	600	3,844	2,690	2,421	2,327	1,111
Total val. exp's.	$2,643	1,305	1,873	484	1,085	884	1,797	1,052

EXPORTS VALUES FOR MONTHS, IN CENTS PER POUND

July	7.1	7.7	13.9	20.7	20.0	27.2	21.2	20.0
August	10.1	6.3	13.5	23.2	22.6	22.8	18.2	18.8
September	11.4	8.0	11.5	25.2	22.9	19.3	28.9	15.3
October	15.5	8.6	9.0	21.2	24.6	16.9	24.2	13.8
November	15.8	10.9	11.6	23.0	24.8	18.0	27.7	13.7
December	16.4	14.3	11.2	22.4	24.7	18.9	20.7	11.1
January	15.6	12.5	10.4	22.4	23.4	18.2	33.8	14.5
February	16.2	12.7	10.5	21.3	23.1	24.1	36.6	14.1
March	13.9	10.7	10.8	21.2	23.0	21.5	32.1	19.4
April	15.0	11.2	8.8	19.0	20.8	26.7	30.4	16.7
May	13.9	11.3	9.6	18.6	22.3	21.1	33.3	17.3
June	14.8	9.3	7.3	16.5	22.4	22.3	30.3	20.5
Average	13.8	10.2	10.6	21.9	23.7	19.2	28.1	14.7
Import val., av.	25.1	20.8	19.1	58.4	40.3	35.2	44.7	16.1

COURSE OF PRICES AT NEW YORK CITY FOR CHOICE STATE HOPS

	1897-8	1896-7	1894-5	1893-4	1892-3	1891-2	1890-1	1889-90
September	8½@14	8½@10	9 @12	22 @24	20 @25	15 @ 18	23 @28	14 @16
October	13 @17	9 @11	9 @11	21 @24	22½@25	16 @ 17	43 @47	11 @13
November	16 @18	10½@15	10 @13	22 @23	22½@25	19 @ 21	35 @47	11 @14
December	17 @18	14 @16	11@12½	21½@23	23 @24	20 @ 22	32 @45	13@14½
January	18@20	14 @15	10½@12	21½@22½	23 @24	21 @ 28	32 @38	13 @16
February	18 @19	13½@14	10 @11	21 @23	23 @25	25 @ 27	33 @36	13 @20
March	17 @19	11½@13	10 @11	18 @21	21 @23	24 @ 25	28 @31	14 @20
April	16½@18	10@11½	9 @10	18 @19	21@21½	26 @ 32	27 @32	16 @18
May	15½@16½	10@10½	8 @10	16 @18	20½@22	28 @ 30	29 @32	18 @20
June	15 @16½	9 @10	8 @9	14 @16	21@22½	24 @ 29	30 @32	20 @22
July	12½@14½	9 @10	8 @9	12 @14	21 @22	24 @ 26	22 @28	20 @23
August	11½@12½	9 @10	7 @10	10 @12	21 @22	24½@27	17 @20	21 @28

HOP CROP OF THE WORLD

Crop of	a 1897	1896	1895	1894	1893	1892	1891	1890
Germany	310	353	368	404	130	300	269	164
Austria	99	136	95	109	74	79	72	65
France	38	43	42	38	33	44	36	54
England	256	281	343	395	257	257	272	176
Total	**703**	**813**	**848**	**946**	**494**	**680**	**649**	**459**
U.S.	200	175	292	320	268	223	208	205
Aggregate	**903**	**988**	**1140**	**1266**	**762**	**903**	**857**	**664**

a American Agriculturist's preliminary estimate. This journal is an accepted authority on America's hop crop, but it frankly admits that this crop is one of the most difficult to report upon for obvious reasons. The figures of each crop are subject to final revision at the close of each year when data are available of the interior and foreign movement.

HOP CROPS AND PRICES

This table shows, for many years, the bales of hops produced each season in the United States and in Europe (including England), the total constituting about 95% of the world's supply. It also gives the number of bales of each crop exported from the United States, and the imports of foreign hops into the

United States, with average yearly United States export prices and Hamburg import values.

c Crop of	U.S. crop.	Europe crop.	Total crop.	U.S. exports.	U.S. imports.	a U.S.	b Hamburg.
	In thousands of bales of 180 lbs. net.					Av. cents. Per lb.	
1897	200	695	895	95	14	13.8	
1896	175	813	988	63	17	10.2	
1895	292	848	1,140	93	15	8.8	
1894	320	946	1,266	97	17	10.7	
1893	268	494	762	97	5	22.0	
1892	223	680	903	63	15	23.7	
1891	208	649	815	70	13	19.3	26
1890	205	459	664	49	21	26.6	26
1889	218	717	935	42	36	29.0	20
1888				69	31	22.4	21
1887				39	28	17.4	19
1886				1	103	21.0	21
1885				76	15	12.5	24
1884				35	9	19.7	33
1883				75	4	24.1	50
1882	125	387	512	43	12	71.8	40
1879	95	379	474	54		26.3	26
1874	110	428	538	17		41.9	41
Av. '81-90	193	746	939			H 50.1	L 19
'85-89	190	773	963			H 20.7	L 19
'81-85	196	720	916			H 50.1	L 25
'76-80	152	647	799			H 30.0	L 21

a Average annual export value (in cents per lb.) of hops shipped from the United States. *b* Average annual value (in cents per lb.) of all hops imported into Hamburg, Germany. H, Highest average annual import value of hops imported into Hamburg, during the period noted; L, lowest. *c* Observe that the year given is that in which the crop was produced.

Hops consumed per bbl. of beer: United States, 1 to 1¼ lbs.; England, 1½ to 2½ lbs.; Germany and elsewhere, ¾ to 1½. A barrel of beer, U.S., contains 32 imperial gallons, or 31 gallons net.

Gross weight of a bale of hops: United States, 185 lbs., legal tare 5 lbs., actual tare 7 to 9 lbs.; foreign hops imported into United States, 350 to 600 lbs. per bale, averaging 430 lbs., with a tare of 14 lbs.

ACREAGE IN HOPS (SO FAR AS ASCERTAINABLE)

	1897	1896	1895	1894	1893	1892	1891	1890
England	50,863	54,207	58,940	59,535	57,564	56,259	56,142	53,961
France	6,122	6,428	7,939	7,264	6,921	6,728	6,592	6,968
Germany	98,767	101,709	103,923	104,241	103,901	107,282	107,791	110,681
Austria	35,108	36,431	39,765	38,048	37,626	36,857	36,679	38,708
Total Europe	**190,860**	**198,785**	**210,567**	**209,088**	**206,012**	**207,126**	**207,204**	**210,318**
New York	19,735	22,190	26,238	30,177	32,300	33,100	34,600	35,000
Washington	3,000	4,500	5,700	10,000	9,000	8,000	6,101	4,338
Oregon	9,000	12,000	16,500	15,000	10,000	6,000	3,900	2,620
California	6,000	7,200	8,500	8,600	8,000	7,000	5,340	4,015
Total United States.	**37,735**	**45,890**	**56,938**	**63,777**	**59,300**	**54,100**	**49,941**	**45,973**
World's aggregate	**228,595**	**244,675**	**267,505**	**272,865**	**265,312**	**261,226**	**257,145**	**256,291**

Yield per acre kiln-cured packed hops: Europe, 400 to 750 lbs., say 500 lbs. in good year; England, 905 lbs. in '97, 936 lbs. in '94 and 888 lbs. as the official average for the years 1886-95; New York, 400 to 1,200 lbs., say 900 lbs.; Pacific coast, 600 to 2,000 lbs., say 1,200 as a fair average in a good year. Germany's ten-year average is 510 lbs. per acre, largest crop averaged 730 lbs. per acre in 1894, lowest 260 lbs. per acre in 1893.

COMPARATIVE RANK OF UNITED STATES HOP SECTIONS, ACCORDING TO THE CENSUS OF 1890

Rank.	Counties.	State.	Acres.	Total crop. lbs.	Per acre. lbs.
1	Otsego	New York	7,749	4,698,687	606
2	Madison	New York	6,956	4,094,440	589
3	Oneida	New York	6,002	3,704,341	617
4	Pierce	Washington	2,191	3,699,671	1,689
5	King	Washington	1,768	3,238,075	1,831
6	Schoharie	New York	5,563	3,148,885	566
7	Sacramento	California	963	2,134,606	2,217
8	Sonoma	California	1,046	1,263,610	1,208
9	Marion	Oregon	974	1,169,657	1,201
10	Franklin	New York	2,930	1,106,123	378

UNITED STATES CENSUS OF HOP CROPS IN POUNDS.

	1890.	1880.	1870.	1860.	1850.
New York	20,063,029	21,628,931	17,558,681	9,671,931	2,536,299
Washington	8,313,280	703,277	6,962	44	0
California	6,547,338	1,444,077	625,064	80	0
Oregon	3,613,726	244,371	9,745	493	8
Wisconsin	428,547	1,966,827	4,630,155	135,587	15,930
Other	205,350	558,895	2,626,062	1,183,587	944,792
Total	**39,171,270**	**26,546,378**	**25,456,669**	**10,991,996**	**3,497,029**

UNITED STATES CENSUS OF ACREAGE AND VALUES

STATES.	Acreage.			Values.	
	1890.	1889.	1879.	1890.	1889.
New York	35,552	36,670	39,072	$6,068,163	$2,210,137
Washington	5,282	5,113	534	2,284,955	841,206
California	3,796	3,974	1,119	1,521,847	605,842
Oregon	3,223	3,130	304	1,047,224	322,700
Wisconsin	871	967	4,439	142,198	51,983
Other states	238	358	332	41,037	27,829
U.S.	48,962	50,212	46,800	$11,105,424	$4,059,697

GERMANY'S FOREIGN TRADE IN HOPS.
IN BALES, 180 POUNDS NET.

Crop of	Exports.	Import.	Net export
1896	111,495	39,103	72,392
1895	135,613	22,647	112,922
1894	41,746	50,615	a 38,869
1893	118,516	20,803	97,713
1892	118,020	24,729	93,219
1891	124,778	14,970	109,808
1890	208,999	22,747	186,252
1889	127,100	13,095	114,005
1888	144,197	15,153	124,044
1887	226,010	13,975	212,035
1886	167,880	19,439	148,441
1885	159,322	13,646	145,676
1884	93,497	19,445	74,052
1883	21,864	7,258	14,606
Average.	129,327	19,656	109,993

a Net import.

GREAT BRITAIN'S IMPORTS OF HOPS BY YEARS. (BALES.)

Cal yr.	From U.S.	Total.	Value U.S.	Value Other.	Export value
1896	76	129	14c	12c	14c
1895	95	130	15c	13c	16c
1894	68	118	20c	18c	21c
1893	88	127	26c	24c	30c
1892	50	117	24c	22c	31c
1891	50	121	23c	22c	35c
1890	45	117	22c	20c	35c
1889	48	125	15c	15c	21c
1888	56	135	17c	16c	

IMPORTS, EXPORTS AND TOTAL SUPPLY IN GREAT BRITAIN.
(THOUSANDS OF BALES, 180 LBS. NET.)

	Exports.	Imports.	Net Import	English crop	Tot. supply
1897				256	
1896	7	129	122	257	380
1895	8	135	127	344	471
1894	13	118	105	396	501
1893	11	127	116	258	374
1892	7	117	110	257	367
1891	6	122	116	272	388
1890	8	117	109	177	286
Av. '90-5	9	123	114	284	398
1889	11	124	113	——	——

U.S. FOREIGN TRADE, SUPPLY AND CONSUMPTION. *A*

Crop of	Exports	Net supply	Imports	Total supply	Con.[1]
1896	63	112	17	129	177
1895	93	199	15	214	184
1894	97	223	17	240	184
1893	97	171	5	176	171
1892	63	170	15	175	178
1891	70	138	13	151	164
1890	49	143	21	164	157

[1] Consumption at 1 lb. of hops to a bbl. of beer. *a* In thousands of bales.

ENGLISH HOP CROPS COMPARED.

Crop of	In thousands bales, 180 lbs.			Av. yearly prices, cts. per lb.	
	Kent.	Oth.	Tot.	Export.	Import.
1897	153	103	256	16	15
1896	199	83	282	14	12
1895	197	147	344	16	13
1894	264	132	396	21	20

Average import value in 1893 was 26c per lb.; '92, 24c; '91, 23c; '90, 22c; '89, 15c; and in 1888 it was 17c.

HALF-YEAR'S FOREIGN TRADE, JAN. 1 TO JULY 1, IN BALES.

Great Britain	1897	1896	1895
Imports	35,206	62,404	65,770
Exports	3,497	3,016	3,450
Net. Imports	31,709	59,388	62,324
United States			
Exports	21,554	42,237	48,881
Imports	8,126	8,507	6,157
Net exports	13,428	33,730	42,734

U.S. HOP CROPS COMPARED. [IN THOUSANDS BALES OF 180 LBS. NET.]

Crop of	Pacific coast.	N.Y. state.	Total U.S.	Av. exp. value lb.
1897	135	65	200	15.4c
1896	100	75	175	10.2c
1895	182	110	292	8.8c
1894	180	140	320	10.7c
1893	143	125	268	22.0c
1892	105	118	223	23.7c
1891	94	114	208	19.3c
1890	92	100	192	26.6c

A ten-year statement (1883-'92) shows Germany's *imports* to have come, on the average, from Austria 90%, from Belgium 3%, from France 1%, from Russia 2%, from other countries 4%. Germany's *exports* during the period noted averaged: To Austria 6%, to Russia 2%, to England 6%, to France 13%, to Belgium 11%, to Sweden 2%, to the United States (average for 1887-'92) 8%, to other countries 24%. Germany's exports to the United States ranged from 7,500 cwt. (of 110 lbs) in 1887 to 46,000 cwt. in the trade year ended Aug. 31, 1890.

WORLD'S PRODUCTION AND CONSUMPTION OF HOPS

For 1884-'86, from the *Deutschen Hopfenbau Verein,* which gave it up in 1889 as unreliable. From 1887-'96, from the Vienna *Brewers' Journal.*
[In metric hundredweight of 110 lbs.]

	World's consumption.	World's production.		World's consumption.	World's production.
1884	——	1,604,400	1893	1,669,791	1,481,300
1885	1,549,000	1,888,550	1894	1,725,762	2,205,510
1886	1,655,000	1,846,810	1895	1,744,439	2,012,155
1887	1,698,026	1,607,000	1896	1,923,756	1,994,370
1888	1,615,000	1,569,200	Total cwt	19,893,128	22,256,700
1889	1,606,486	1,967,250	Av.	1,657,760	1,712,054
1890	1,546,915	1,096,000	Av. bales 180 lbs.	1,013,075	1,046,255
1891	1,566,642	1,456,440			
1892	1,592,311	1,527,715			

BEER PRODUCTION

United States data are official. Returns for the world are from *Brewers' Journal,* Vienna. [In millions of barrels of 31 gallons.]

Year	U.S.	World.	Year.	U.S.	World.
1865	4	——	1891	30	167
1870	7	——	1892	32	172
1880	13	——	1893	35	173
1885	19	——	1894	33	175
1887	23	147	1895	33	179
1888	25	145	1896	35	197
1889	25	148	1897	34	200
1890	28	166			

RELATIVE CONSUMPTION OF BEERS AND OTHER LIQUORS

E. Struve, in *Wochenschrift für Brauerei,* Berlin, 1896, estimates that the wine consumed in France, Germany, Switzerland and Belgium contains 6% of alcohol, against 7% in Austria, Holland, Denmark, Sweden and Norway, and 8% in Great Britain, Russia and the United States. For the beer consumed an average proportion of 4% alcohol was adopted, and for spirituous liquors 33.3%. The table is based on official returns and shows the annual consumption per capita for 1895.

In liters (1 liter equals 2.113 pints U. S., or just about 1 quart).

COUNTRY	Wine	Beer	Spirits	Alcohol consumption			
				Wine	Beer	Spirits	Total
Belgium	3.7	169.2	14.1	0.22	6.76	4.7	11.68
France	103.0	22.4	12.4	6.18	0.90	4.04	11.12
Denmark	1.0	33.3	26.7	0.07	1.33	8.9	10.30
Germany	5.7	106.8	13.2	0.34	4.27	4.4	9.01
Great Britain	1.7	145.0	8.4	0.13	5.80	2.8	8.73
Switzerland	55.0	37.5	9.3	3.30	1.50	3.1	7.90
Aus.-Hungary	22.1	35.0	12.4	1.54	1.40	4.15	7.09
Holland	2.6	29.0	14.1	0.18	1.16	4.7	6.14
Russia	3.3	4.7	14.1	0.26	0.19	4.7	5.15
Norway	1.0	15.3	12.0	0.07	0.61	4.0	4.68
United States	1.8	47.0	7.7	0.14	1.88	2.58	4.60
Sweden	0.4	11.0	4.8	0.03	0.44	1.6	2.07

BELGIUM HOP TRADE, 1890

	Lbs.	Value.
Exports	12,111,228	$2,443,546
Imports	10,586,895	2,533,636
Net exports	**1,524,333**	

QUOTATIONS ON HOPS

In America, are in cents per lb. avoirdupois. In England, are in pounds and shillings per cwt. of 112 lbs. In Germany, are in marks per metric cwt. of 110 lbs. Reckoning one mark as equal to 23.8c, and £1 (one pound sterling) at

$4.86, the following tables show the equivalent of foreign quotations in U.S. currency per lb.

German Marks per 110 lbs. equal U.S. cents per lb.

5M = 1.08c	35M = 7.59c	50M = 10.81c	80M = 17.30c
10M = 2.16c	38M = 8.22c	54M = 11.68c	90M = 19.47c
20M = 4.32c	40M = 8.65c	60M = 12.98c	95M = 20.55c
25M = 5.40c	42M = 9.08c	68M = 14.70c	100M = 21.36c
30 M = 6.49c	45M = 9.73c	70M = 15.14c	

English pounds and shillings per 112 lbs. equal U.S. cents per lb.

1£ = 4.33c	1£ 8s = 5.59c	1£ 18s = 8.18c	5£ = 21.65c
1£ 1s = 4.54c	1£ 10s = 6.49c	2£ = 8.66c	6£ = 25.98c
1£ 2s = 4.75c	1£ 12s = 6.90c	3£ = 12.99c	7£ = 30.31c
1£ 3s = 4.91c	1£ 14s = 7.33c	4£ = 17.32c	8£ = 34.64c
1£ 5s = 5.40c	1£ 15s = 7.57c		

THE GERMAN HOP GROWERS' ASSOCIATION

Is quite an effective institution, under the presidency of Herr Von Soden with Mr. A. Fairth as vice president. Its prime object is the obtaining of crop reports from its branch associations and local members, as well as the dissemination of information of general interest to planters. Its official organ is *Deutschen Hopfenbau-Verein,* edited by Mr. Fairth, to whom we are indebted for numerous courtesies. The branch organizations of this association, with the director of each and his post-office address, are as follows:

BAVARIA—Spalt and Spalterland, director Landrath Herkenschlager, Hauslach bei Georgenogmund; Hersbruckerland, director T. von Soden, Vorra; Neustadt, director Steward Sorg, Newstadt on the A; Oberbayern, director Mayor of Aichbichlerland and Delegate Imperial Diet, Wolnzach; Niederbayern, director G. Zieglmeier, Katzenhofen; Kinding, director Burgomaster Zaigler, Kinding.

WURTEMBERG—Neckar, director Verwalter Distlen, Flemmingen; Schwarzwalkdkreis, director City Counsel Edelmann, Rottenburg on N; Donankreis, director Delegate of Diet Bueble, Tettnang.

BADEN—Director Burgomaster Mechling, Schwetzinger.

EAST AND WEST PRUSSIA—Director Wiepkingin, Tathannen.

The bulk of the trade in hops in Germany is concentrated at Nuremberg, but every large town has trade chambers, at which there is some buying of hops.

The principal market places for Austrian hops are Saaz, Auscha and Danba. In Alsace, Haguenan is the most important hop center; Frankfort-on-the-Main, Mayence-Mannheim are also important centers of great hop dealers, but not for the sale of hops by growers. For Baden, the hop center is Schwetzinger. In Belgium, Alost and Poperinghe are the chief centers, but the dealers also meet regularly at the exchanges in Brussels and Antwerp, where large transactions are conducted. In France the most important hop markets are Dijon and Lunéville.

THE HOP DICTIONARY

Glossary of Technical Terms Pertaining to Hops and the Hop Trade,
Including Hints on Curing and Other Practical Points

BY N. F. WALTER

GROWERS who have not good equipment (facilities in one locality may not suffice in another) and cannot get an experienced and good dryer, should never contract their hops because

CURING is difficult; the variety, nature, development of the hop, and climatic and weather conditions playing an important part. Each season they may differ, and often each day, and even picking (morning and afternoon) must be differently manipulated. Proper knowledge of drying takes years to acquire and cannot be learned in several seasons.

HARVESTING, HANDLING, CONDITION—Hops should be well cultivated and cared for, be of good, bright, even color, well matured but not over-ripe, cleanly picked and properly cured; put up in sound condition and merchantable shape, in new material and in correct and uniform bales.

CONTRACTS call for choice goods, unless distinctly otherwise specified; that is, an excellent article and only the best grade. Usage has established that all contracts—purchases and sales—are made "severally as to bales," and there is no averaging to grade or sample, each bale stands on its own merits. There is no averaging to a grade or sample, because buyers often have absolutely no outlet for anything below the standard bought on, and further, that below certain grades there is often positively no market, and it is therefore impossible at times to estimate the value of inferior goods.

INSPECTION—Buyer has the right of accepting each and every bale equal to purchase, and the privilege of rejecting all that class below. Changes in quality, due to difference in bulk, wrapping, storage and general outward surroundings and conditions, may take place in a few hours, and therefore samples may not represent the distinct bales from which they were taken. Example: A sample taken from hops, newly baled in slack condition, wrapped in paper and mailed, might dry out, and reach intending purchaser entirely changed, whereas a redrawn (a fresh sample) or a tryer sample, from the same bale, would show such slackness and would therefore be reason to reject. In other words, the identical bale, the original sample from which reached the buyer with every indication of soundness, would, due to large bulk, depending upon storage conditions largely, either heat or sour.

For the reason cited in foregoing example and other changes that might occur, it is an accepted fact that samples, as a rule, are unreliable as an indication of the condition (and therefore quality) of a hop in bale, except at the time they are drawn. Changes often take place so quickly, as already explained, that samples are often useless as a guide a few days, and even at times a few hours, after they are taken.

JUDGMENT—The reason that the inspection by buyer, or his expert agent, is, through usage, accepted as final, is because there must be some experienced judge to determine whether the hops are up to requirements and in sound and proper condition, and the buyer, after acceptance, assumes all risks in changes that may take place in storage and transit, while seller is relieved from all responsibility after he has delivered.

Experts, through varied experience, can tell the merits and defects of a hop, and may be able to attribute the cause of shortcomings, though they are rarely hop dryers.

DISEASE or VERMIN, SPRAYING and WEATHER DAMAGE are always good causes for rejection under contracts.

STORAGE—Hops, being delicate and sensitive, should have clean, good, cool, dry and dark storage—removed from moisture and away from anything that emits a decided or strong odor.

WEIGHTS—Only full pounds count on each bale. Hops lose in weight, with age.

GRADING—In the trade there are four divisions made in quality: First, choice; second, prime; third, medium, and fourth, common to poor.

QUALITY AND CLASSIFICATION DO NOT CHANGE, BUT COMPARATIVE VALUES DO—Examples: A hop grading prime remains a prime hop, although in an excited market it may command the same price as choice, whereas, in a weak market, it is rarely worth more than medium. In years of world's shortage, common and poor bring as much as medium. In such seasons, there are really only two distinctions in the price of brewing hops, choice and prime commanding about the same figure, while there is little if any difference between medium, common and poor. In years of overproduction, choice only command top price, prime less, and medium sink to the lowest valuation; it is difficult to get a price even on mediums, when, of course, common and poor are practically valueless. Summing up: Supply and demand regulate values, but do not alter quality.

HOP LAW is principally "Law Merchant." Most terms and expressions in the hop business are purely technical. Rulings and decisions governing quality, condition, samples, inspection. etc., in cases covering other kinds of crops, will not necessarily apply to hops, because of their very nature, which is so different and distinct from other products. The customs controlling the tender, delivery, inspection, rejection, replacing and acceptance, are well established, and therefore controlled by trade usages, that is "Law Merchant."

THE HOP GLOSSARY

ACRE—A hop acre is sometimes figured, regardless of land surface, at 1000 hills, or plant centers, as an acre, but unless so qualified means statute acre.

AGING—Becoming old; or taking on the properties of former years' growths. Losing in brewing virtue.

AIRING—Permitting a free circulation of air between the bales or through the hops. When found slackish, the bales are separated "on end," to retard damage that would be promoted by close piling. The bales are sometimes opened at the side seams, and the hops loosened and holes made through them, to permit access of air, thus to prevent or arrest heating. Often the bales are torn apart and the hops opened out and spread on the cooling and kiln floors, to permit free exposure to the air. These should really be subjected to re-drying. Hops that require airing are generally sour, and therefore poor. The injury to value naturally depends upon the degree of resultant damage to quality and market conditions.

ALL FAULTS—The English term for "as is."

AS IS—Without privilege of rejection. A condition at times imposed on sales of lots containing damaged hops, or including injured and unmerchantable bales or a mixed lot.

BABY BALE—A single, small, "lightweight" in another wise standard weight lot.

BABY HOPS—See young hops.

BAD COLOR—See off color.

BAKED—Harsh and rough, with brewing quality damaged through improper ventilation and circulation of air in kilns during drying process.

BALES—The packages of hops as they are marketed. These must be of regulation shape and requirements. See baling; also weight.

BALING AND DIMENSIONS—Size varies somewhat, but the hop presses generally in use are the recognized standards. The baling must be properly done, neat and clean, with new material and well sewed. See weight.

BALINGS—See pickings.

BASKETS—Small baskets used in gathering hops.

BATCH—A single kiln flooring of hops.

BERRY OR STROBILE—The hop. The catkins. Infrequently called buds (not burrs).

BIN—The cooling room, or the divisions of a cooling room, which is sometimes partitioned off into compartments, to keep qualities (pickings, color and curings) separate for proper baling. A name also given to a burlap bottomed framework into which green hops are picked.

BINSMAN—In England, the person in charge of a gang, who also pulls poles and assists measuring and loading wagons. See field boss; also poleman.

BLACK—A name sometimes given to heated hops.

BLEACHING—The sulphuring. The term bleached is applied to hops that are naturally whitish, or those that have lost too much color from improper sulphuring. See sulphuring.

BLIGHTED—Diseased.

BOLD—Rather large and prominently flaky hops, that are serviceable but not silky.

BOARDS—A term used in England for the shelves or tables on which the hop samples are shown.

BOARDY—Hops hard pressed and wanting in life; not springy in bale.

BOOKER—In England, the person who follows the measurer and enters proper credits or gives tickets to each picker for work done.

BOXES—Boxes in which the picked hops are taken to the kilns.

BOX MAN—See foreman.

BRACTS—See petals.

BREADY—The aroma of warm, newly baked bread. An indication of over-drying. See over-drying.

BRIGHT—Brilliant and even in color.

BROKEN—The berries parted, the petals largely loose and showing few whole berries, resulting from too much drying, untimely or improper baling. See also hard pressed; also shelly, powdered and chaffy.

BUDDING—A reprehensible method of throwing selected whole berries on the face of samples.

BUDS—See catkins; also berry.

BULK SAMPLES—A big representative line of samples from a lot.

BURNT—See roasted.

BURR—The burr or real bud—the undeveloped hop in its early stages of formation, before the petals form.

BUTTERY—See oily.

BUTTONY—Full berried. See flaky.

BUYING AND SELLING—The purchase and sale of hops, which is always according to grade or sample. Unless otherwise distinctly agreed, the transactions are subject to usual customs.

CABBAGY—The cut edge of a sample resembling a cut section of a cabbage. Also called streaky.

CAKED—Brick-like, lumpy (the berries sticking together in bale and lifeless). Indicates slackness. See also cold.

CARLOAD—Unless otherwise qualified, means sixty regulation bales. See bales, baling, weights.

CARPELS—See petals.

CARPET—See kiln cloth.

CASING (or going through case in curing)—This is a reactionary sweating that takes place in the cooling room, promoting mellowness or silkiness, sometimes named the "second sweat," calling the "reek" the first sweat.

CATKINS OR BUDS—See berry; for buds, see also burr.

CHAFFY—Broken and brittle, powdery; also called mashy.

CHEESY—The name sometimes given to the rancid odor of hops that are aging. See aging.

CHIPPY—See cold; also harsh.

CHOICE HOP—One that shows the entire bale to be of a good, bright, even color, flaky (whole berries), cleanly picked, silky, rich in lupulin, good flavor and properly cured and baled. See quality.

CLAMMY—A cold, moist, sticky condition, indicating slackness. See slack.

CLASSIFY—To grade samples. See grading.

CLEAN—Refers to picking. Free of leaves and stems and all foreign matter. Properly picked.

CLOTH—See hop cloth; also kiln cloth.

COARSE—Full berries, wanting in silkiness.

COLD—Clammy, slackish and usually immature. Also applied to a hop that has soured. If the hops do not heat, but simply sour in bale, then the berries will be hard and slightly caked, or what may be termed chippy. See souring.

COMMON HOP—One that shows either flaky berries, or, if broken (not powdered), fair brewing quality. It may be somewhat poor in color and general conditions, but must be sound and put up properly. See quality.

COMPLEXION—Color and sightliness, particularly as to luster.

CONDITION—The name sometimes given to lupulin. Brewing virtue. Further, it refers to curing and marketable manner of baling.

CONES—The hops. See berry.

CONTRACTING—The engaging ahead of production. The selling and buying of "future" crops. See buying and selling.

COOKED—See stewed.

COOLING ROOM—The room or building in which the hops, after being taken from the kilns, undergo the completion of the curing process.

COPPER HOPS—Hops for the brewing kettles. Generally applied to those hops used for boiling, but not particularly adapted to flavoring in the vats, or for hopping.

CORE OR HOP STEM—The strig. The axis of the strobile in contradistinction to the vine stems. See stems.

CROSS GRAINED—Diagonally packed, through tramping, or the "follower" of the press not working evenly. See curly.

CRUSTED—Hardening of the outward portions of the bale next to the cloth, caused by damp storage, and causing caking, discoloring and a musty flavor.

CULLS OR CULLING—Bales rejected upon inspection.

CURING—The process of drying hops on the kilns, together with their management in the cooling rooms. See drying; also curing guides, and casing.

CURING GUIDES—There are no positive rules to regulate the drying of hops. Each flooring may require different manipulation, and this requires study, aptitude and years of practical experience, because seasons differ. The following points, however, are essential. LOW HEAT, GOOD DRAFTS, PROPER SULPHURING, and to complete the curing, intelligent management in the cooling room. See drying; also casing.

CURLY—Applied to a flaky hop when cross-grained.

CUT—The cut side of a sample.

DEAD, DULL OR LEADY—Wanting in color, grayish and of a lifeless character.

DELICATE—Tender in texture and of fine flavor.

DIMENSIONS—See baling.

DIRTY—Refers to picking, and is applied to samples that contain either or both leaves and stems. That is, that the hops were uncleanly, not properly picked.

DISEASED—A choice hop must be absolutely sound. The slightest touch of disease of any kind prevents a hop being classed choice. Even to be classed prime, there cannot be more than the slightest trace of disease present. A medium hop can only contain very little mold; where mold abounds, the hop can not be better than common and is rarely classed above poor. In seasons where disease is isolated, and the world's crop fairly abundant, then sound hops only are marketable, and those unfortunates in the infected districts can claim no standing for such of their crop that shows disease. It is worthless in a commercial sense, and no buyer takes the product unless bought on sample. See quality.

DOUBLE BALE—The compression of two bales into one package. See repacking.

DRIFTS—The different blocks or portions of a field of hops allotted to separate sets, companies, gang, division or section of pickers.

DRYER—The man who dries the hops, supposed to be an adept at curing.

DRYING—That portion of the curing process that takes place in the kilns.

DULL—See dead.

EARLY HOPS—Early varieties, or early ripening hops.

EDGE—The cut side of samples. Also those parts of a bale where any two surfaces meet. Sometimes applied to the front of the bale; which is properly called the face.

ENDS—Those parts of a bale making top or head surfaces.

EXAMINATION—The examining of a lot of hops and the passing on samples, but this does not constitute inspection unless each bale is tried. See inspection; also good light.

EXCELLENT BREWING HOPS—Prime hops. See quality; also prime.

EXPERT—One versed in the quality and action of hops, together with the requirements and customs of the trade.

EXPORT—Fit for export shipment.

FACE—Surface of the sample; also the long, narrow surface of the bale. See facing; also edge.

FACING—The method sometimes employed of cleaning leaves and stems from the face of samples.

FAIR BREWING HOPS—Common hops. See quality; also common.

FALSE PACKED—Layers of different color or maturity in one bale, or different qualities of hops baled together.

FANCY BREWING HOPS—Choice hops. See quality, also choice.

FAT—The term applied to such hops, during curing, in which the cores have not perceptibly started to dry. See also rich.

FEATHERY—Fluffy. Berries on the vine that are not full or firm, and on which the petals spread.

FIELD BOSS—The overseer of the pickers.

FIRST YEAR'S GROWTH—See first year's planting.

FIRST YEAR'S PLANTING—Also called first year's growth. See young hops; also new.

FLAKY—Lying in layers of full, whole berries. Perfect strobiles.

FLAT—Lifeless and often scrubby, and at times cabbagy.

FLAVOR—See good flavor.

FLOORING—The quantity of hops put on a kiln at a single drying. Depths of floorings cannot be fixed. This depends upon the nature of the hops, capacity, or rather efficiency, of the kilns and the prevailing atmospheric conditions. The character of the hops and possible draft must govern quantity.

FORCED DRYING—Dried too rapidly at a higher than necessary, but not scorching heat; causing loss of volatile oils and making the hops harsh. See harsh.

FOREMEN—The yard bosses, paymasters and superintendents of the different work. These include in the eastern states the sacker, who empties the eight-bushel boxes of green hops each into a separate sack and gives pay checks. On the Pacific coast the box man who supervises the picking and gives pay tickets. The chief baler who is in charge of the presses. See also dryer, weigher, field boss, measurer and booker.

FOXY—Reddish brown from over-maturity, disease or decay.

GOING—Occasionally applied to hops that are heating. See heating.

GOING OFF—Getting overripe, or beginning to show disease. See shattering.

GONE—Applied to hops that have heated. See heating.

GOOD BREWING HOPS—Medium hops. See quality, also medium.

GOOD COLOR—A brilliant pale green, or a golden yellow. A light, bright, properly and evenly developed appearance.

GOOD FLAVOR—The natural aroma emitted from a rich and mature, perfect, healthy hop berry at the time it is taken from the vine, after compression and rubbing between the fingers, and which flavor should permeate the fresh cured hops.

GOOD LIGHT—Hops must be examined by daylight; artificial light will not answer. A soft, clear, steel light. A veiled or slightly shaded (not shadowed) natural light, or what might be styled "an indirect sunlight" is best. The intense direct rays of the sun are too strong. See examination; also inspection.

GOOD QUALITY—See quality.

GRADE AND VALUE—There are four mercantile divisions, namely, choice, prime, medium (or good brewing) and common to poor. Classification does not change, but supply and demand regulate relative values.

GRADING—Classifying hops as to quality and standard.

GRAIN—See cross-grained: also curly.

GROSS WEIGHT—The entire weight of bales. See weights.

GUIDES—See curing guides.

GUMMY—Resinous and not thoroughly cured.

HARD PRESSED—Too heavily baled. It takes buoyancy, or sponginess, from a well-cured hop. Also called hard baled and heavy baled, and when over the maximum proper weight range of 205 pounds are called over-weights, or heavy weights, which latter term is also often applied to a lot of bales, the most of which run considerably over the ordinary average. Sec weights.

HARSH—Wanting in oily matter. Generally flaky, and the berries stiff, or what could be called chippy. Usually the result of forced drying.

HEAD—See top.

HEATING AND HEATED—Spoiled from heating. Rotting or rotted; usually due to insufficient drying; also possible from becoming wet through absorption of a large quantity of moisture, from exposure, and subsequent sweating with consequent turning and rotting. A heated hop is worthless.

HEATED BERRIES—Berries heated before being put on the kilns.

HEAVY BALED—See hard pressed.

HEAVY WEIGHTS—See hard pressed.

HIGH-DRIED—A degree of curing between over-dried and over-fired, which causes the running of the lupulin and loss of oily matter and brewing strength. High-dried hops are of a chaffy nature.

HIGH-FIRED—See over-firing.

HOP CLOTH—The burlap covering for the bales. This should be 24-ounce cloth, and there must be not more than is necessary to properly cover the bales.

HOPPERS—Hop pickers; those who pluck the hops from the vines.

HOPPING—The occasional practice of putting a few whole berries in barrels of malt liquor.

HOPPING OUT—The transitory stage from burr to cone, or formation of the true hop.

HOP SACKS—The sacks in which the hops are taken to the kilns. The size varies in different localities, but should not be made to hold over 80 pounds green hops, which would require a sack about 60 inches long by 40 inches wide. See pokes; also under "foremen."

HOP STEMS—See core.

HOP YARD—Hop garden, or field of hops.

IMMATURE—Not sufficiently ripe; indicated by green appearance of berry, and pale color of hop seed which when fully ripe is dark purple.

IN CASE—See casing.

IN HOP—Fully contoured strobiles in first stages of development of hop.

INSPECTION—The trying and examination of every bale, and the passing on each bale separately. See examination; also good light, and try; tryer samples.

IN THE SWEAT—See casing.

KILN—The house in which the hops are dried.

KILN BOSS—The man in charge of the curing. See dryer.

KILN CARPET—See kiln cloth.

KILN CLOTH—The covering of the kiln floor, usually 10-ounce burlap. Also called kiln carpet.

LATE HOPS—Used in contradistinction to early hops, where a grower has several varieties (including some earlies), or different fields that mature at differently advanced stages of the harvest season.

LEADY—See dead.

LEAVES—The leaves of the hop vine.

LIGHT—See thin; also good light.

LIGHT-BALED—See loose pressed.

LIGHT WEIGHTS—See loose pressed.

LIVELY—See spongy.

LOOSE BALED—See loose pressed.

LOOSE PRESSED—Not tightly enough baled. Well-cured hops, put up this way, and hops that are aging, lose weight quickly. Also called loose baled and light baled, and, when of less than proper minimum weight range of 170 pounds, are styled light weights, a term also applied to a lot of bales, a good many of which run under the customary average. See weights.

LOT—A number of bales collectively.

LUPULIN—The bitter, buttery, globular secretion in the hops—their principal virtue. In its normal condition, in healthy, properly handled, cured new hops, it has its natural brilliant lemon-yellow color and oily characteristics.

MANAGEMENT—The treatment of the hops. Curing.

MASHY—See chaffy.

MATURE—The stage of development proper for picking. Not immature and not over-ripe.

MEASURER—In English usage, the person who measures the green hops with a bushel basket into the pokes. See booker.

MEATY—A fat, wholesome, but not delicate hop.

MEDIUM HOP—A hop of good brewing quality, that may not be as even in color, or may not excel in distinct qualifications as the better grades, but must be bright, fairly clean, safely cured and well put up. See quality.

MELLOWNESS—Silkiness. See sulphuring; also casing.

MERCHANTABLE—Sound and properly put up. Refers to both contents and covering of package, or to the bale itself.

MIX—A disadvisable habit some growers have of blending their growth in cooling room.

MIXED LOT—A lot containing bales of different qualities.

MIXED COLOR—A mixture of early and late pickings, brought about by a deliberate and thorough mixing of the greener with the more mature hops. See uneven color; also off color.

MIXED CURING—See uneven curing.

MOTTLED—Mixed in color. Green, ripe, and over-ripe or wind-whipped berries baled together indiscriminately, usually the fault of uneven ripening.

MOVE—Changing or tossing hops from one place in cooling room to another.

MUDDY—Imperfectly developed unsightly hops full of dirt and sand.

MUSTY—Foul odor. Fustiness.

NET WEIGHT—The weight of the bales less tare allowance. See weights.

NEW—Hops of the latest crop in contradistinction to old hops. As it gets near to a harvest the term applies also to the growing crop. The spirit of its use and plausible intent govern the meaning. First year's growths are sometimes called new hops, to distinguish them from the product of roots that have produced before. See young hops.

OASTS or OAST HOUSES—The English term for the drying houses. See kilns.

OFF COLOR—Not bright; unsightly. Also used when the color is uneven from any cause. See dull.

OILY or BUTTERY—Sometimes applied to hops that are fat and silky. See silky, also rich.

OLD HOPS—All growths except the latest harvest. (See yearlings, also olds and old olds.)

OLD OLDS—A general term for hops over two seasons old. (Beyond two years removed from the latest harvest.)

OLDS—Hops two seasons old. (Growth of the second year removed from latest harvest.)

ORANGED—The lupulin changed from its original yellow to a deeper or orange color, the effect of imperfect curing, or result of aging. See lupulin.

ORDINARY BREWING HOPS—Poor hops. See quality; also poor.

OVER-DRYING—The drying of a hop for too long a period at a low heat. It lessens brewing strength; that is, it lessens its intrinsic value. It makes a hop tender. Over-drying is less damaging in its effects than high-drying or over-firing.

OVER-FIRING, as the word implies, means excess firing in curing process. It causes more or less evaporation of the volatile oils, causes crystallization of the lupulin and spoils the flavor. Its degree can only be determined by an expert (and this can be said of all faults). In its worst stages it is called burnt or scorched; in the lesser degrees, over-fired or high-dried. General over-firing may mean practically entire loss of brewing quality.

OVER-GRADING—Classing hops at too high a quality. Overrating standard. See grading.

OVER-RIPE—Over-mature; shown by the hops turning red, and in advanced stages by fluffiness or lack of solidity in the berries on the vines.

OVER WEIGHTS—See hard pressed.

PACKAGE—A bale.

PACKERS—See pack hops.

PACKET HOPS—Hops put up for domestic uses in tightly compressed quarter, half and one-pound paper packages. See pack hops.

PACK HOPS OR PACKERS—Hops for packets for druggists' and grocers' trade. See packet hops.

PARCEL—A collective number of bales. See lot.

PETALS—The leaflets of the hop; that is, the carpels or bracts of the strobile.

PICK—Privilege of taking any portions of a lot, subject to usual inspection conditions.

PICKINGS OR BALINGS—When the several portions of a yard are picked in different stages of maturity, the hops are kept separate accordingly in early and late, or early, middle and late balings (or pickings), as is necessary.

PLATTY—The development of a yard unevenly; that is, in blocks, or maturing irregularly, but evenly in separate plats.

POCKETS—An English term for bales, or, rather, pressed bags of cured hops, of weights varying according to locality.

POINTING OR TIPPING—The shriveling of the extreme ends of the point petals, and breaking off of these tips when the hops have reached their fullest development. This feature, with dark-purple color of the seed, indicates, under normal conditions, ideal maturity.

POLEMAN—The person delegated to pull vine poles for pickers, or, in the trellis yard, to get down—by means of a hook and blade attachment to a long scantling—such portions of vine and hops as cling to trellis wires when the vines are pulled down for picking.

POKES—The English term for their hop sacks, into each of which 10 bushels of green hops are put.

POOR COLOR—See off color.

POOR HOP—Any hop having some brewing virtue, but off in general appearance and conditions. It must be sufficiently sound to stand shipment, and although it may lack in color and strength, must be well baled. See quality.

POWDERED—Pulverized. This occurs in baling hops that have been killed on the kilns by extreme high-drying or over-firing. Hops become chaffy and powder as they age or disintegrate, but favorable storage retards this. See chaffy.

PRIME HOP—One having most of the characteristics of a choice, but lacking in some point that does not affect its other general conditions. For instance, a prime hop may be choice other than to be not quite as good in flavor, or not fully rich in lupulin, or not quite although fairly cleanly picked, or the berry instead of being firm may be tender, or the color may not be quite even, though fairly uniform (not far off nor mixed), etc. That is, some slight and single defect, but otherwise equal to choice. A hop sample containing several blemishes as above cited would, as a rule, grade only medium. See quality.

PRIMROSE—An expression rarely used, but sometimes applied to a color indicating early stages of over-ripeness.

QUALITY—To secure good quality, diligent cultivation and attention in the yard, and clean picking, proper curing and baling are essential. See choice, or what could be called fancy brewing hops; prime, that might be termed excellent brewing hops; medium, that are also called good brewing hops; common, or fair brewing hops; and poor, or ordinary brewing hops. Also see diseased, worthless, curing guides.

RANK—An off colored hop with a strong, earthy, or green vegetable aroma.

RE-BALED—Baled a second time. Necessary, of course, to re-dried hops. Also to hops that have been opened out for airing. Sometimes resorted to when the first baling was too heavy or too light, or the baling unmerchantable. Re-baling generally badly breaks the hop. See re-dried; also airing.

RED—A name sometimes given to over-ripe hops.

RE-DRIED—Dried a second time. Administered to hops that are found slack, or to hops that have become wet. A re-dried hop cannot be a choice hop, and the operation generally results in very inferior quality. See re-baled.

REEK—The vapor or steam arising from hops at the early stages of drying, called at times a sweat. See sweating, steaming and casing.

RE-PACKING—The re-wrapping or replacing of the burlap on the bale, or the compression of several bales into one package, a method sometimes employed for exporting.

RICH OR FAT—Thick in lupulin. See fat.

ROASTED OR BURNT—Terms sometimes applied to badly scorched hops.

ROUGH—Unevenly developed berries with bracts lacking in oil and not smooth.

RUB—See the rub.

RULES—See curing guides.

BUSTED—Brown spots. A weather effect on delicate points and flaws of the growing hops. A blemish, but this is not meant as the disease known as rust.

SACKER—See foremen.

SACKS—See hop sacks.

SAFE—Sufficiently dry in bale to stand either long (including export) shipment or compact piling and close storage.

SAMPLES—A chunk (or, as it is called, a square sample) of hops, cut and drawn from side of bale, with knife and tongs. Advance samples, the type or shipping samples. Type samples, the standard for comparison of quality. Shipping samples, those sent to indicate style of hops shipped. Re-drawn samples, fresh or newly drawn samples, to show nature and condition. Samples should not be taken until the hops have set or become firm in bale, which takes several days after pressing. At inspection examination a square sample is usually taken from each 10th to 15th bale, depending upon size of lot, besides the tryer samples from each bale. See tryer samples; also bulk samples.

SCORCHED—Burning, caused by over-firing, resulting in the crystallization of the lupulin and excessive loss of and injury to brewing quality and flavor.

SCRUBBY—Lacking in roundness or fullness of berry, wanting in solidity. Light, flat berries.

SECOND SWEAT—See casing.

SEEDLESS—Free or almost free of seed.

SELECTION—The separate accepting or rejecting of each bale severally in a crop of hops.

SELLING—See buying and selling.

SHATTERING—The breaking apart of the berry. Falling off of the petals. Usual to hops that have gone off. Also through excessive drying-out in the bin after casing. See going off; also shelly.

SHELLY—Brittle from drying out in cooling room. Rather shattered in baling. See casing; also shattering.

SHIPPERS—The top quality demanded by foreign trade.

SHOVE OFF—Act of removing the dried hops from the kiln floors.

SICKLY—Cured hops showing an unhealthy or darkened, watery discoloration of the base of the petals and of the lupulin, caused through faulty handling.

SIDE—The broadest and largest surface of a bale. The samples are drawn from this portion.

SILKY—Oily feeling in a hop. A point indicating proper development and good condition.

SIZE—See baling.

SKYLIGHTS—The windows through which the light is reflected on the boards.

SLACK BINE—Shortage of lateral arms and foliage.

SLACK-DRIED or SLACKNESS—See slack.

SLACK OR UNDER-DRYING—A slack hop is one that is under-dried, not sufficiently dry. Hops in this condition heat or sour, depending on the degree of slackness and storage conditions. Heating, which virtually means rotting, may be occasioned by even a bunch as little as a handful of slack hops, and once started, usually affects the entire bale, and even communicates to surrounding bales if closely stored. See heating or heated; also sour and cold.

SLACKISH—Slightly slack. These hops usually sour. See slack.

SLACK-SCORCHED—Hops dried at too high a temperature and not dried through. Burning them without drying them, so that while they have a scorched flavor, they may still sour or heat in cooling bin or bale. Often the fault of too many hops being on the kiln for its capacity. See slack.

SLEAZY—Thin in texture, wanting in vitality and brewing strength; flimsy.

SMOKY—Smoky smell that the hops take when the kiln furnace or pipes are defective and allow smoke to escape through the hops.

SMOTHERED—Inappropriate casing, resulting from inadequate airing of hops in cooling room. Usually due to too heavy packing, causing loss of brilliancy and effecting early disintegration or aging.

SMUDGED—Incipient heating arrested. Berries indicating that they had started to heat and cooled off.

SOFT—Delicate to the eye and touch, and usually mild of flavor. Yielding to easy pressure.

SOGGY—Very wet or slack. See slack.

SOUND—Not slack; in a safe condition.

SOURING—A hop generally sours or takes a sour flavor when not properly ventilated in cooling rooms (when too heavily piled), while going through case, and generally, if properly dried, when baled too soon; that is, before completely cured. A slackish hop in bale will sour if it does not heat. Condition of storage affects the extent of damage at times. See storage; also cold.

SPINDLING—Thin, straggling, light-foliaged, unproductive or small bearing vines.

SPONGY—Springy to the touch, full of life. A good point and essential to a first-class hop. Also called lively.

SPOTTED—Berry showing uneven development of color. Petals of different color in same berry.

SPRAYING—Squirting washes on the vines through spraying machines, to avert the appearance or stop the spread of vermin or disease.

SPRAYING DAMAGE—Injury created by the spraying wash, caused generally by washes that are too powerful, or by applying same at wrong stage.

STEAMING—Emitting volumes of moisture. This occurs to the hops in the kiln at the earlier stages of drying. See the reek; also sweating.

STEMS—The twigs from the lateral arms (consisting of the peduncle, petioles and pedicels), which should not be picked. See core, or hop stem.

STEWED OR COOKED—A condition due to inadequate overhead drafts in kilns, causing the moisture supercharged air or reek to fall back on the drying hops.

STIR—Some growers stir instead of turning their hops by walking through or rather dragging their feet with a shuffling motion along the kiln floors through the batches of drying hops.

STORAGE—Should be clean, dry and dark, away from moisture and foreign odors. Hops while in transit are in a poor condition of storage, due to the extreme and oppressive heat generated in cars and vessels.

STOUT—Rich in lupulin and of good flavor.

STOWAGE—An English expression for cooling room.

STRAWY—The cut edge of a sample of broken or scrubby hops that lack in oily matter, and have a straw-like appearance.

STREAKY—See cabbagy.

STRIPPING—The removing of the foliage (branches and leaves) from the lower portion of the main vine.

STROBILE—See berry.

STRONG—Full flavor.

SULPHURING—Burning sulphur at the kiln furnaces, so that the fumes pass through the drying hops. It has a three-fold effect; it opens the hops, thus helping to keep them loose, which assists the draft; it modifies or evens the color of the hops; and it has a preserving tendency by promoting mellowness. See casing. It should be employed at the proper stage, and that is from the time the hops on the kiln have become warm until they have finished steaming. See bleaching.

SUN SCALD—The weakening of the vine and injury to or curtailment of the crop, through protracted intense heat during the developing period of growth, before the true hops have formed.

SWEATING—Sometimes used in the sense of heating. Often for the reek. Also applied to casing, which is at times called the second sweat.

SWEEPINGS—The refuse from floors swept into the press, making such bales inferior.

TARE—The customary allowance or deduction in weight for baling cloth.

TENDER—Soft; delicate to the touch.

THE RUB—The feeling and action of a hop between the fingers or hands in examination.

THICKNESS—An English term employed in passing on the quantity of lupulin; richness.

THIN—Lacking in lupulin; wanting in brewing strength. Also called light, or weak.

TINTED—Touched with a faint pinkish color, indicating the turning point to over-maturing. This feature is desirable. An indistinct blush, not too pronounced, or it would mean over-ripeness.

TIPPING—See pointing.

TONGS—A tool for taking square samples from a bale.

TOP CROP—The growth of hops running principally to the extreme end (top) of the vines, due to less than ordinary branching or arming, and indicating a lighter than normal yield.

TOP OR HEAD—The smallest surface of a bale.

TOUGH—A tenacious condition that the cores of the hops are in at a certain stage of curing. A number of tough stems in baled hops are an indication of slackness.

TOUGH STEMS—Tenacious "hop stems;" strigs that are not brittle; incompletely cured cores.

TRAMPING—The light compression of the hops in the presses, to permit more hops being added for proper weight of bale before actual power is applied through the follower attachment of the press.

TRY—The probing with tryer. The examination of each bale singly. See examination, also inspection, and good light.

TRYER—A harpoon-shaped instrument used in inspecting each bale, and which brings out a handful of hops. See try.

TRYER SAMPLES OR TRYINGS—The handful of hops taken from the center and sometimes from several parts of each bale, with the tryer, by the inspector. See sample, also good light, and inspection.

TRYINGS—See tryer samples.

TURN OR TURNING—Some growers upset or turn their hops on the kiln floor after several hours' drying. A hop is said to be turning when aging. Also said of hops that are heating or heated. See souring.

UNDER-DRYING—See slack.

UNDER-GRADING—Underrating quality. Classifying below proper standard. See grading.

UNEVEN COLOR—Not a uniform color; a mixture of differently colored but fairly developed berries. See mixed color; also off color.

UNEVEN CURING AND MIXED CURING—Uneven drying of hops, caused by too heavy floorings; that is, too great a depth of hops on kilns, or through faulty kiln construction and improper drafts, so that in order to dry part of the hops properly those in another section of the kiln are either over or under-dried. In such cases it is uneven curing. Where the kilns work properly and growers dry some floorings to different degrees than others and mix them in bin and bale it is mixed curing.

UNSAFE—See unsound.

UNSOUND OR UNSAFE—Not sound: slack or slackish.

USEFUL—Not particularly fine, nor sightly, but of good brewing quality.

VALUE—See grade and value.

VARIEGATED—Mixed in color; checkered. See mottled.

VERMIN DAMAGE—Injury to the growing crop, caused by pests and the resulting damage of which is apparent in the hop.

WEAK—See thin.

WEIGHER—The yard boss, who has charge of pickers, weighs the hops and gives credit, or pay-checks, for them.

WEIGHTS—Bales should weigh from 170 to 205 pounds gross weight and should average not less than 180 pounds net weight. Unless otherwise stated all transactions imply net weight. See loose pressed, hard pressed, bales, baling.

WIND-WHIPPED—The tips and outer leaves of berries bruised, withered and discolored, caused by wind shaking and hitting.

WOODY—Abounding in vegetable fiber and harsh.

WORTHLESS—Hops that cannot even be classed poor; that is, those that are spelled through bad handling or disease. There is always a lot of this valueless trash. See quality.

YARD BOSSES—Those in charge of the picking. See foremen.

YEARLINGS—Hops of the next to the latest harvest. (On the Pacific coast young hops are sometimes erroneously called yearlings.)

YOUNG HOPS—Hops of first year's planting; i.e., from vines of the first growth after the sets or cuttings have been planted for a crop; infrequently called baby hops. See new.

RULES REGULATING THE HOP TRADE

AMONG MEMBERS OF THE NEW YORK PRODUCE EXCHANGE.
[Adopted March 1, 1883, and amended September 27, 1889.]

Rule 1.—At the first meeting of the Board of Managers, after their election, the president shall (subject to the approval of the Board) appoint as a committee on hops, five members of the New York Produce Exchange, who are known to be dealing in hops, to consist of two brewers and three dealers. It shall be the duty of this committee to properly discharge the obligations imposed upon them by these rules, and also to consider and decide all disputes arising between members dealing in, consuming, or exporting hops, which may be submitted to them.

A majority of the committee shall constitute a quorum, but the committee shall fill temporary vacancies, if requested by either party, by some member or members representing the same interest as the absent member or members, and a decision of a majority of those present at any meeting shall be final. They shall keep a record of their proceedings, and a fee of fifteen dollars ($15) shall be paid to the committee for each reference case heard by them—to be paid by the party adjudged to be in fault, unless otherwise ordered by the committee; provided, however, that nothing herein shall prevent a settlement of questions of difference by private arbitration, or as provided in the by-laws.

Rule 2.—All transactions in American hops only between members of the Produce Exchange shall be governed by the following rules, but nothing herein shall be construed as interfering in any way with the right of members to make such special contracts or conditions as they may desire.

Rule 3.—All hops shall be deliverable in merchantable bales. When a certain number of pounds are sold, number of bales not specified net weight shall be understood.

Rule 4.—When specific lots are sold by sample, or otherwise, and are ready for immediate delivery, any bale weighing not less than 170 pounds, nor more than 205 pounds, shall be considered a good delivery.

Rule 5.—When hops are sold for future delivery, and the weights of the bales have not been ascertained at the time of sale, a good delivery shall be a sufficient number of bales to effect a delivery of the number of bales sold, at an average of not less than 180 pounds, nor more than 190 pounds, gross weight.

Rule 6.—On all New York state hops, an allowance of five pounds per bale shall be made as tare, in conformity with Chapter 239, laws of 1889.

Rule 7.—In the absence of any specific agreement, the seller shall have the right to demand payment at the time of passing the title.

Rule 8.—Whenever sales are made between members of the Produce Exchange through a broker who is not a member of the exchange, a written memorandum of the transaction is to be exchanged by the principals before the sale is binding.

Rule 9.—Hops sold for immediate delivery must be inspected on the day succeeding the sale. Hops sold for future delivery must be inspected on the day succeeding the notice of delivery.

Rule 10.—If upon inspection it shall be found that any lot, or part of a lot, of hops shall not conform with the contract, the buyer shall take all which do conform to the contract, and the seller shall replace the lot, or part of a lot, rejected with other hops of as good a quality, and for this purpose the seller shall have 10 days to replace and tender hops to fill the original contract; but if a specific lot is sold by sample the buyer shall take all which are up to sample, and he shall have the privilege of taking the rejections at a reduction to be agreed upon between seller and buyer, or to be settled by arbitration.

Rule 11.—Hops shall be weighed (unless otherwise agreed upon) by a city weigher, whose return shall be taken as the correct weights of the bales. Weigher's fees to be divided by buyer and seller equally.

Rule 12.—All hops shall be removed at the buyer's expense within two days after receiving the invoice (weather permitting), and until then the seller is to hold the same fully covered by insurance at invoice value.

Rule 13.—When hops are sold to arrive and to be inspected on dock, the buyer shall, after inspection and order for delivery being given, assume the same relation toward the transportation line by which the hops arrived, as the seller previously held as regards their removal from the place of delivery within the time granted by such lines for that purpose.

Rule 14.—Rules 3, 4, 5 and 6 shall apply only to the crop of 1883 and subsequent crops.

Rule 15.—A carload of hops shall be understood to contain not less than 10,000, or more than 13,000 pounds.

BIBLIOGRAPHY

The Curiosities of Ale and Beer, John Bickerdyke; London, 1889.

Twenty-five Years of Brewing and History of American Beer, George Ehret; New York, 1891.

The London and Country Brewer, printed for T. Astley; London, 1758-59.

Hops and Hopping, John B. Marsh; London, 1892.

Root Growing and the Cultivation of Hops, Arthur Roland, edited by William H. Ablett; London, 1887.

Hops, their Cultivation, Commerce and Uses in Various Countries; a manual of reference for the grower, dealer and brewer, P. L. Simmonds; London, 1877.

Hop Culture in the United States, being a practical treatise on hop growing in Washington territory, from the cutting to the bale, E. Meeker; Puyallup, Wash., 1883.

Diseases of Plants Induced by Cryptogamic Parasites, Tubeuf and Smith; London, 1897.

Diseases of Plants, H. Marshall Ward (contains a detailed popular description of mildew); New York, 1890.

Insects and Fungous Enemies of the Hop Vine, Charles Whitehead; Journal of the Royal Society of England series 3, 1893, pp. 240-247.

Methods of Preventing and Checking Attacks of Insects and Fungi, Charles Whitehead; London, 1891.

Annual Report of the Board of Agriculture, Charles Whitehead; London, 1890, p. 24.

Hop Cultivation, Charles Whitehead; London, 1893.

Handbook for Hop Growers, a guide to the practical culture of hops (German), E. D. Strebel; Stuttgart, Germany, 1887.

Hop Cultivation (German), C. Beckenhaupt; Weissenburg, Germany, 1890.

A beautiful set of photographic plates, each 36 by 23 ctms., was prepared in Germany by Dr. M. Braungart, and published in 1881-2. Not less than 429 "varieties" of hops are illustrated in the 37 plates. The hops are shown in life-size and from all parts of the world. The old edition is out of print, but copies are in many libraries in Europe. A new edition is expected in 1901 that ought to be in every agricultural college and other important libraries in America.

Hop Culture, practical details as given by ten experienced cultivators residing in the hop-growing sections in the United States, collected by Orange Judd Company, edited by A. S. Fuller; New York, 1883.

www.ingramcontent.com/pod-product-compliance
Lightning Source LLC
Chambersburg PA
CBHW031843200326
41597CB00012B/249